建筑立场系列丛书 No.36

体验与感受
艺术画廊与剧院

Experiencing and Sensing
Art Gallery and Theater

中文版
（韩语版第352期）

韩国C3出版公社 | 编

王凤霞　于风军　蒋丽 | 译

大连理工大学出版社

4 建筑脉动
学校设施和公共领域

004 公共领域、体育场所和学校之间的连接点 _ Andrew Tang
006 坎贝尔体育中心 _ Steven Holl Architects
016 马德里自治大学的马约尔广场 _ MTM Arquitectos

32 体验与感受

艺术画廊与剧院

032 体验艺术画廊与剧院 _ Heidi Saarinen
038 奥克兰美术馆 _ Francis-Jones Morehen Thorp + Archimedia
048 柏林建筑图纸博物馆 _ Speech Tchoban & Kuznetsov
056 水田艺术博物馆 _ Studio Sumo
066 让·克洛德·卡里埃剧院 _ A+ Architecture
076 乌镇大剧院 _ Artech Architects
088 艺术馆 _ Future Architecture Thinking
094 树美术馆 _ Daipu Architects
106 丰岛横尾馆 _ Yuko Nagayama & Associates
116 普图伊演艺中心 _ Enota

建筑立场系列丛书　No.36

124 邻里住宅
立面和内部世界

124 通往未知世界的大门 _ Marta González Antón
128 幸运书屋 _ Chang Architects
138 Binh Thanh住宅 _ Vo Trong Nghia Architects + Sanuki + Nishizawa Architects
146 Venturini住宅 _ Adamo-Faiden
152 Alcobaça住宅 _ Aires Mateus
164 103住宅 _ Marlene Uldschmidt Architects
174 Flynn马厩改建住宅 _ Lorcan O'Herlihy Architects

180 建筑师索引

Archipulse
University Facility and Public Realm

004 *Connecting Dots between Public Realm, Sports and School _ Andrew Tang*

006 Campbell Sports Center _ Steven Holl Architects

016 Plaza Major of UAM _ MTM Arquitectos

Experiencing and Sensing
Art Gallery and Theater

032 *Experiencing Art Gallery and Theater Buildings _ Heidi Saarinen*

038 Auckland Art Gallery _ Francis-Jones Morehen Thorp + Archimedia

048 Museum for Architecture Drawings in Berlin _ Speech Tchoban & Kuznetsov

056 Mizuta Museum of Art _ Studio Sumo

066 Jean-Claude Carrière Theater _ A+ Architecture

076 Wuzhen Theater _ Artech Architects

088 House of the Arts _ Future Architecture Thinking

094 Tree Art Museum _ Daipu Architects

106 Teshima Yokoo House _ Yuko Nagayama & Associates

116 Ptuj Performance Center _ Enota

Dwell How
The Facade and the Inner World

124 *Gate to an Unexpected World _ Marta González Antón*

128 Lucky Shophouse _ Chang Architects

138 Binh Thanh House _ Vo Trong Nghia Architects + Sanuki + Nishizawa Architects

146 Venturini House _ Adamo-Faiden

152 Alcobaça House _ Aires Mateus

164 103 House _ Marlene Uldschmidt Architects

174 Flynn Mews House _ Lorcan O'Herlihy Architects

180 Index

学校设施和公共领域

公共领域、体育场所和学校之间的连接点

大学不仅仅总是关于教室、课本和因拖沓的学习而导致的整夜苦读。许多学生对大学的美好印象来自于其他的活动,如聚会、娱乐和体育活动。娱乐中心作为学生们聚会、交流、会面和单独学习的中心,迎合了这些学生们的需要,同时又提供设施供他们锻炼并保持身体健康。因为学术性运动员既想要锻炼身体又渴望有学术上的进步,学校强调在该中心内设计一些基础设施,以满足所有学生的需求。新设施的侧重点不仅仅是体育场、更衣室和负重训练室。在这里,身体和心灵成为队友,都在其共同成功的道路上发挥同样重要的作用。

因为许多学校都位于城市环境中,娱乐中心正在渐渐成为学校的一个发挥代表作用的地方,成为人们由临近社区或公共交通运输路线接近时第一眼看见的建筑。这些设施通常位于校园的边缘并起到连接周围社区、公共空间和学校私人项目区的交通的城市集合点作用。其环境旨在为建筑创造了一个"移动"和与周边相应的机会:如纽约哥伦比亚大学的坎贝尔体育中心那样悬吊于空中,以产生一个欢迎入口,同时俯瞰高架火车;或像马德里自治大学的马约尔广场建筑那样深入景观设计,以产生聚集人群和用桥梁连接周围建筑的开放公共空间。明白了协同的重要性,世界各地的建筑师们正在营造这些公共集会空间,以促进学生们的成长。

MTM建筑事务所设计的马德里自治大学的马约尔广场建筑坐落于原有的林荫大道的尽头,它为学校的新项目提供场地。该场地形成

Connecting Dots between Public Realm, Sports and School

A university has not always been about classrooms, textbooks and procrastination induced all-nighters. Many students' fond memories of college come from other activities such as social gatherings, recreation and sport events. Recreation Centers cater to these students as hubs for students to gather, exchange, meet and study individually while offering facilities to exercise and condition their bodies. As more demands are put on scholar-athletes to perform physically as well as academically, schools have emphasized providing a supportive infrastructure within which all students can succeed. New facilities are focusing on more than just sport fields, lockers, and weight rooms. The body and mind have become teammates, each as important as the other in ensuring their collective success.

As many schools are situated in urban settings, recreation centers are becoming a representative part of the school, being the first building to be seen when approaching from adjacent neighborhoods or public transportation lines. These facilities are usually at the edge of the campus and serve as urban anchors to connect the surrounding neighborhoods, public spaces and transportation to the private programs of the schools. Its contextual ambition creates an opportunity for the architecture to "move" and respond to its surrounding: lifting into the air to create a welcoming portal while overlooking the elevated trains as is the case of Campbell Sports Center of Columbia University in New York; or digging into the landscape to create open public space to gather and bridge with surrounding buildings as is the case with the Plaza Major Building of The Autonomous University of Madrid (UAM). Understanding the importance of collaboration, architects worldwide are helping to create these communal gathering spaces for students to grow.

The Plaza Major building at UAM by MTM Arquitectos in Madrid is situated at the end of an existing boulevard, housing the school's new programs. It offers a healthy mix of both private and public spaces, together forming a communal open space. With its lack

University Facility

坎贝尔体育中心/Steven Holl Architects
马德里自治大学的马约尔广场/MTM Arquitectos

公共领域、体育场所和学校之间的连接点
/Andrew Tang

了私人与公共空间的有益组合,两种空间共同组成了一处公共开放空间。由于没有界定持续拓展的边缘,场地的空间布局和设计明显地反映了协作中包容性的重要性。七座桥可以通往广场内部,使内外形成流畅的过渡,而马蹄铁形的屋顶位于景观之上12m,作为林荫大道的扩建区。屋顶除了提供室外娱乐空间之外,因光伏电池板排放于屋顶边界,它也为设施提供能量来源。整个综合实施和周边地区的协同作用得到了充分的发挥。随着教室和娱乐空间的精心设置和混合,能源的公用和协同作用就很容易实现了。而在太平洋的彼岸,建筑在提供一处共同发展的、成功的空间方面发挥着相同的作用。

坐落于曼哈顿北部的坎贝尔体育中心发挥着从繁忙城市街道到Bakers综合体育中心的门户的作用。学生们可以通过迈进和穿梭于新中心来实现与新环境产生的身体和精神上的过渡。鉴于"空间线、平面点"的设计理念,新建筑想要表达运动场在团队沟通中起到的图表作用,提议其应该在空中移动姿势,以形成一个欢迎参观者进入体育场的入口,同时也表达出斯蒂文•霍尔雕塑品质的特点。该建筑是一个体现心灵和身体的三层不同规划的结构,每一层都体现其中的一种主要目的:服务于身体的体力和健身层;服务于心灵的学习空间和接待区;服务于心灵/身体的学生运动员剧院。这两个项目中学习和娱乐的协同作用体现学术结构对待努力学习和尽情玩乐之间的一种平衡态度。或许这两个项目的动态性暗示我们应该尽情地玩乐。

Campbell Sports Center/Steven Holl Architects
Plaza Major of UAM/MTM Arquitectos

Connecting Dots between Public Realm, Sports and School
/Andrew Tang

of defined edges allowing for continuous movement, the spatial layout of the site and design clearly reflect the importance of inclusivity in collaboration. Seven bridges provide access to the interior of the plaza, allowing a seamless transition from the outside in, while the roof, shaped like a horseshoe, is elevated 12 meters above the landscape and serves as an extension of the boulevard. The roof, in addition to providing outdoor areas for recreation, is a source of the facilities energy as PV panels line the perimeter. The synergy of the entire complex and surrounding area is on full display. With classrooms and recreational spaces deliberately placed and mixed throughout, communal and collaborative energies are effortless. Just across the Atlantic, architecture has played a similar role in providing a space for communal growth and success.
Nestled comfortably in Northern Manhattan, The Campbell Sports Center serves as a gateway from the bustling city streets to the existing Bakers Athletic Complex. Students are able to make a physical and mental transition into their new surroundings by stepping into and through the new center. With a design concept of "lines in space, points on the ground," the new building conveys the field play diagrams used in team communication, that suggests the building making moving gestures into the air to create a portal that welcomes visitors into the playing fields while also delivering a sculptural quality characteristic of Steven Holl. The building is stratified in three layers of different programs reflecting the mind and body, with each layer featuring one of the main aims: The strength and conditioning level for the body; the study space and hospitality zone for the mind; the student athlete theater for the mind/body. The collaboration of study and recreation for success in both of these projects reflects the academic institution's attitude towards a balance of working hard and playing hard. Perhaps the dynamism of both projects implies that we should play harder.

Andrew Tang

and Public Realm

坎贝尔体育中心
Steven Holl Architects

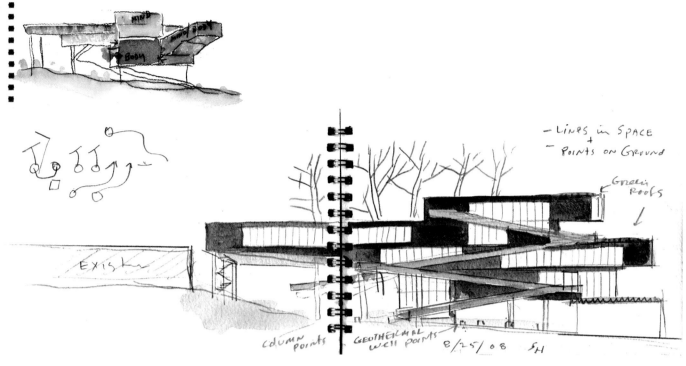

坎贝尔体育中心项目位于西218号街和百老汇的拐角处——曼哈顿的最北端,此处百老汇与第十大道和地铁一号线的高架轨道相交,该中心形成了一个通往贝克田径综合楼的新"大门"。贝克田径综合楼是哥伦比亚大学户外体育项目的主要竞技设施。

自20世纪70年代中期所建的马塞勒斯·哈特利·道奇体育健身中心以来,坎贝尔体育中心是哥伦比亚大学校园内的第一个新建的田径场,它将成为贝克田径综合楼新的活力,并为整个校际竞技项目提供更多的项目空间。该中心增加了约4459m²的空间,设有力量训练和健身训练空间、各项运动活动室、剧院式会客室、用于接待的多功能套间和学生–运动员学习室。

坎贝尔体育中心旨在为有抱负的学生–运动员提供心灵、体魄和心灵/体魄方面的服务。

其设计理念"平面点、空间线"从斜坡场地上的某个地基点发展。这就像足球、橄榄球和棒球的场地布局图一样,其中的点和线模拟比赛场地上的推拉动作。

该建筑塑造了百老汇和218号街的一个城市街角,然后向上抬升形成一个入口,将体育场与街景连接起来。蓝色的挑檐在阶梯的景观中得以延伸,强调了贝克田径综合楼具有城市规模的门廊的开放性。

作为"空间线",该建筑的平台和外部楼梯使体育场可以向上攀行进入大楼,而人们从上层建筑可以欣赏到体育场甚至曼哈顿的景色。

该建筑采用清水混凝土和钢结构,以及磨砂铝立面,使得该建筑与贝克体育场的独特历史建立起一种连接。1693年,横跨斯派特·代夫尔河的国王大桥是通往曼哈顿的主要通道。百老汇大桥目前的基础设施以数百吨的起重能力架起了高架地铁和百老汇。而这些细节和结构都体现在了坎贝尔体育中心项目当中。

Campbell Sports Center

Located on the corner of West 218th Street and Broadway – the northernmost edge of Manhattan, where Broadway crosses with Tenth Avenue and the elevated tracks of the 1 Subway Line – the Campbell Sports Center forms a new gateway to the Baker Athletics Complex, the primary athletics facility for the Columbia University's outdoor sports program.

It's the first new athletics building to be constructed in Columbia University's campus since the Marcellus Hartley Dodge Physical Fitness Center was built in the mid-1970s, the Campbell Sports Center will be the new cornerstone of the revitalized Baker Ath-

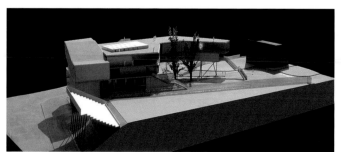

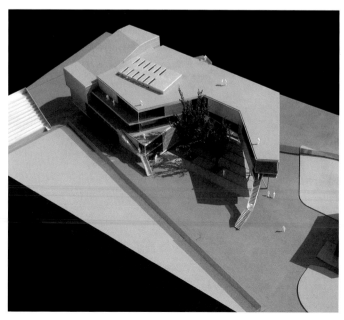

letics Complex and provides increased program space for the entire intercollegiate athletics program. The facility, which adds approximately 48,000 square foot of space, houses strength and conditioning spaces, offices for varsity sports, theater-style meeting rooms, a hospitality suite and student-athlete study rooms.

The Campbell Sports Center aims at serving the mind, the body and the mind/body for aspiring scholar-athletes.

The design concept "points on the ground, lines in space" – like field play diagrams used for football, soccer, and baseball – develops from point foundations on the sloping site. Just as points and lines in diagrams yield the physical push and pull on the field, the building's elevations push and pull in space.

The building shapes an urban corner on Broadway and 218th Street, then lifts up to form a portal, connecting the playing field with the streetscape. Extending over a stepped landscape, blue soffits heighten the openness of the urban scale portico to the Baker Athletics Complex. Terraces and external stairs, which serve as "lines in space," draw the field play onto and into the building and give views from the upper levels over the field and Manhattan.

With an exposed concrete and steel structure and a sanded aluminum facade, the building connects back to Baker Field's unique history. In 1693, The Kings Bridge, which spanned the Spuyten Duyvil Creek, was the main access rout into Manhattan. The current infrastructure of the Broadway Bridge carries the elevated subway, and Broadway, with a lift capacity of hundreds of tons. Its detail and structure are reflected in the Campbell Sports Center.

项目名称：Campbell Sports Center
地点：New York, NY, United States
建筑师：Steven Holl, Chris McVoy
主要合伙人：Chris McVoy
项目副主管：Olaf Schmidt
项目团队：Marcus Carter, Christiane Deptolla, Peter Englaender, Runar Halldorsson, Jackie Luk, Filipe Taboada, Dimitra Tsachrelia, Ebbie Wisecarver
施工经理：Structuretone
结构工程师：Robert Silman Associates
M.E.P.工程师：ICOR Associates
土木工程师：Hirani Engineering
可持续性工程师：Transsolar
幕墙设计顾问：W.J. Higgins
照明工程师：Wald Studio
音像顾问：The Clarient Group
声学顾问：Cerami Associates
甲方：Columbia University
总建筑面积：4,459m²
设计时间：2008 竣工时间：2013
摄影师：©Iwan Baan (courtesy of the architect)

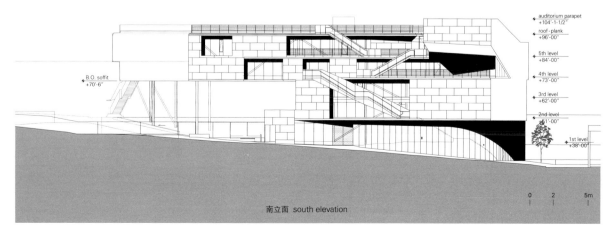

南立面 south elevation

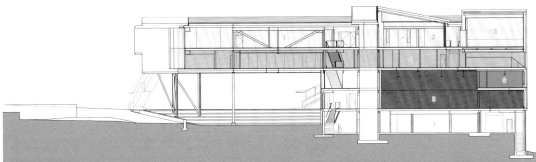

1 学生-运动员学习中心 2 待客套间 3 剧院式会客室 4 足球套间 5 力量训练和健身训练空间 6 大学教练套间 7 场地维修区
1. student-athlete study center 2. hospitality suite 3. theater style meeting room 4. football suite 5. strength and conditioning space 6. varsity coaches suite 7. field maintenance space

A-A' 剖面图 section A-A'

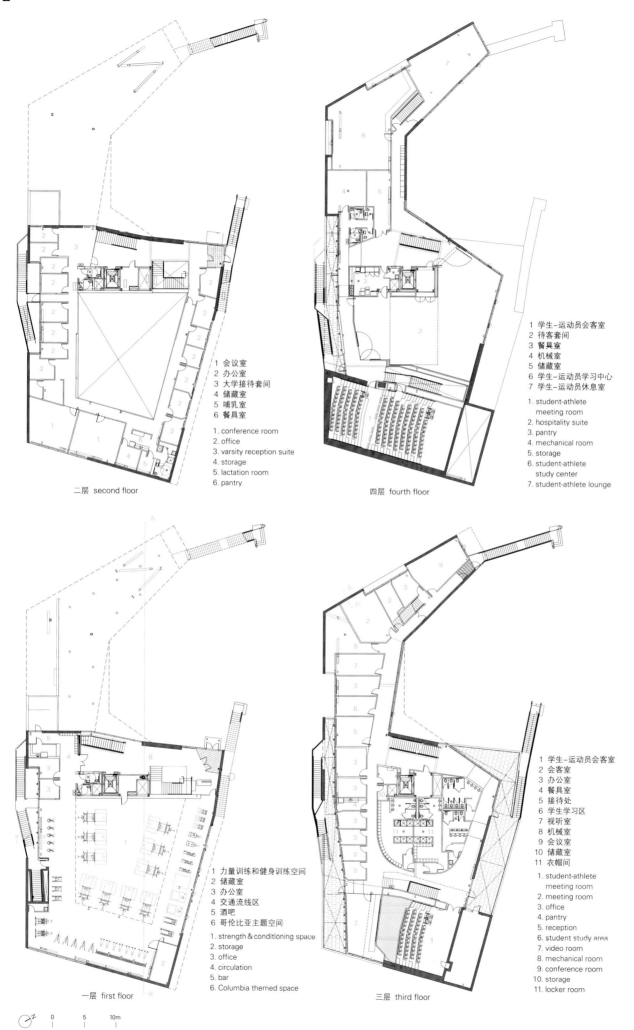

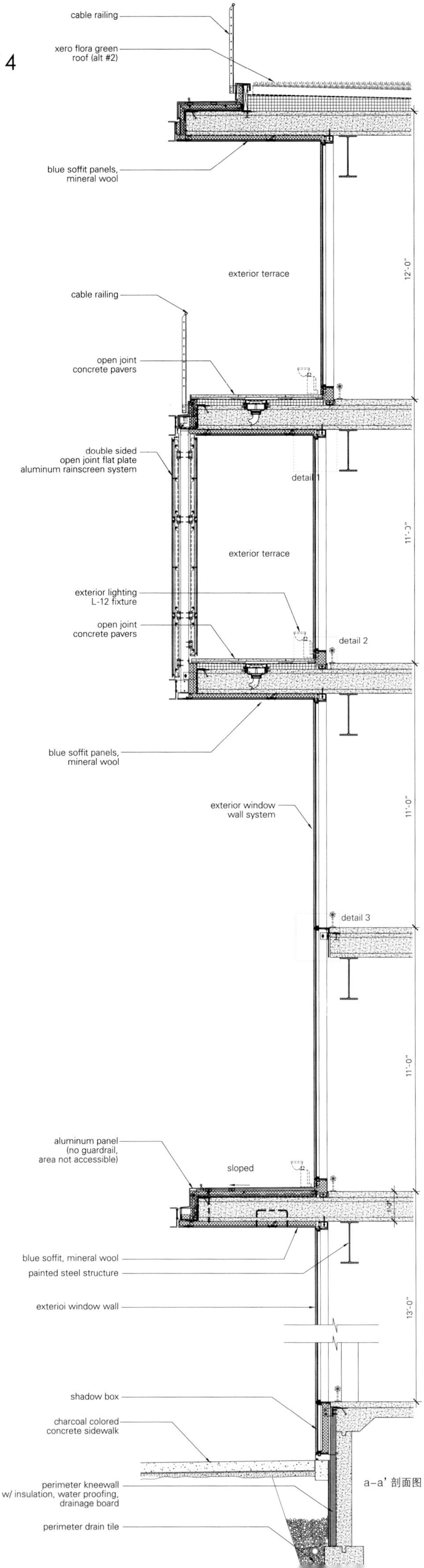

Section a-a' labels (left drawing)

- cable railing
- xero flora green roof (alt #2)
- blue soffit panels, mineral wool
- exterior terrace — 12'-0"
- cable railing
- open joint concrete pavers
- double sided open joint flat plate aluminum rainscreen system
- detail 1
- exterior terrace — 11'-0"
- exterior lighting L-12 fixture
- open joint concrete pavers
- detail 2
- blue soffit panels, mineral wool
- exterior window wall system — 11'-0"
- detail 3
- aluminum panel (no guardrail, area not accessible)
- sloped
- blue soffit, mineral wool
- painted steel structure
- exterior window wall — 13'-0"
- shadow box
- charcoal colored concrete sidewalk
- perimeter kneewall w/ insulation, water proofing, drainage board
- perimeter drain tile

a-a' 剖面图 section a-a'

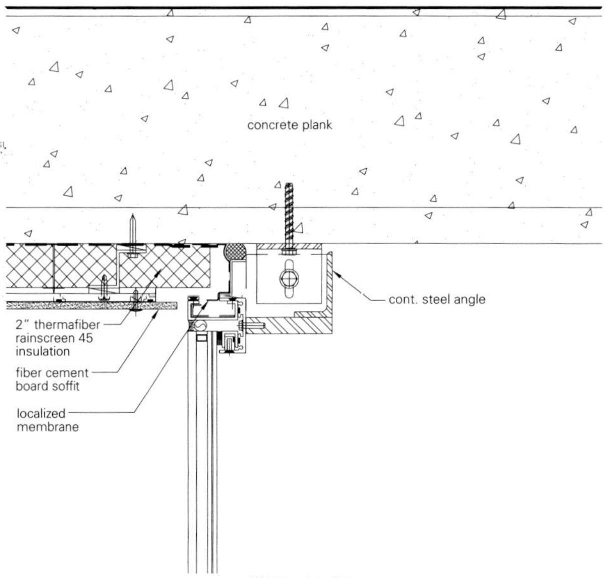

详图1 detail 1

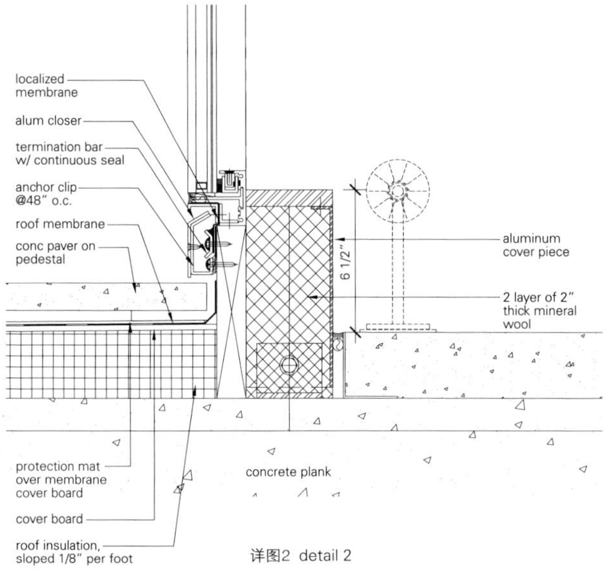

- concrete plank
- cont. steel angle
- 2" thermafiber rainscreen 45 insulation
- fiber cement board soffit
- localized membrane

详图2 detail 2

- localized membrane
- alum closer
- termination bar w/ continuous seal
- anchor clip @48" o.c.
- roof membrane
- conc paver on pedestal
- aluminum cover piece
- 6 1/2"
- 2 layer of 2" thick mineral wool
- protection mat over membrane cover board
- cover board
- concrete plank
- roof insulation, sloped 1/8" per foot

详图3 detail 3

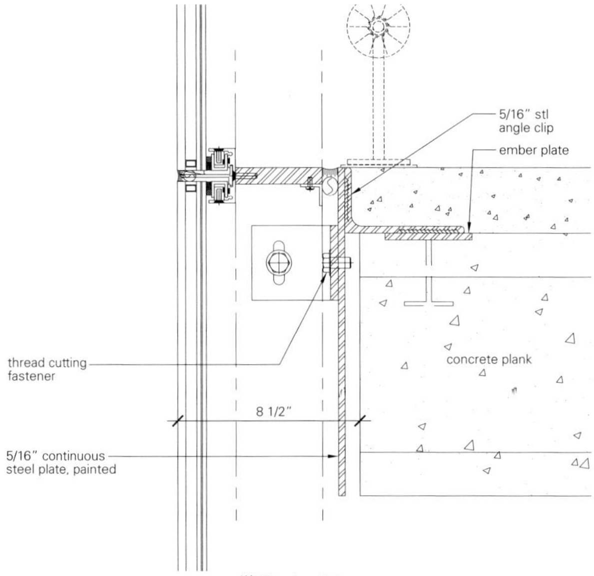

- 5/16" stl angle clip
- ember plate
- concrete plank
- thread cutting fastener
- 8 1/2"
- 5/16" continuous steel plate, painted

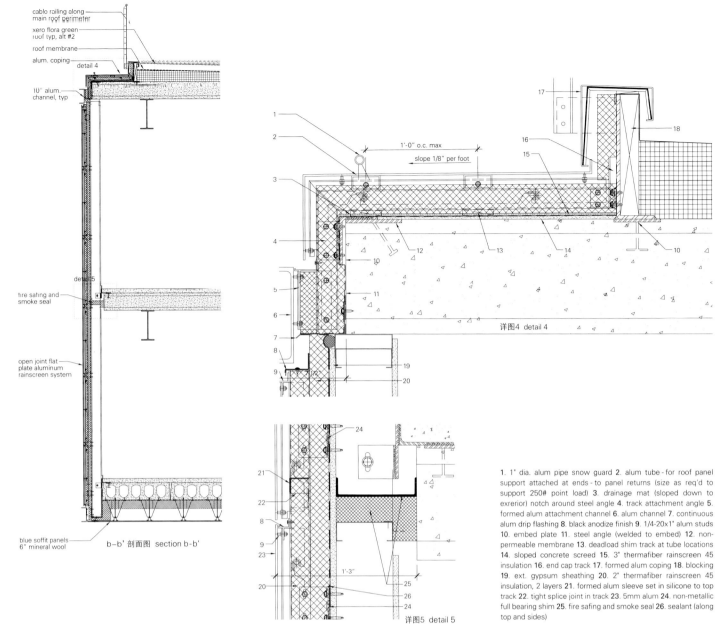

1. 1" dia. alum pipe snow guard 2. alum tube - for roof panel support attached at ends - to panel returns (size as req'd to support 250# point load) 3. drainage mat (sloped down to exrerior) notch around steel angle 4. track attachment angle 5. formed alum attachment channel 6. alum channel 7. continuous alum drip flashing 8. black anodize finish 9. 1/4-20x1" alum studs 10. embed plate 11. steel angle (welded to embed) 12. non-permeable membrane 13. deadload shim track at tube locations 14. sloped concrete screed 15. 3" thermafiber rainscreen 45 insulation 16. end cap track 17. formed alum coping 18. blocking 19. ext. gypsum sheathing 20. 2" thermafiber rainscreen 45 insulation, 2 layers 21. formed alum sleeve set in silicone to top track 22. tight splice joint in track 23. 5mm alum 24. non-metallic full bearing shim 25. fire safing and smoke seal 26. sealant (along top and sides)

马德里自治大学的马约尔广场
MTM Arquitectos

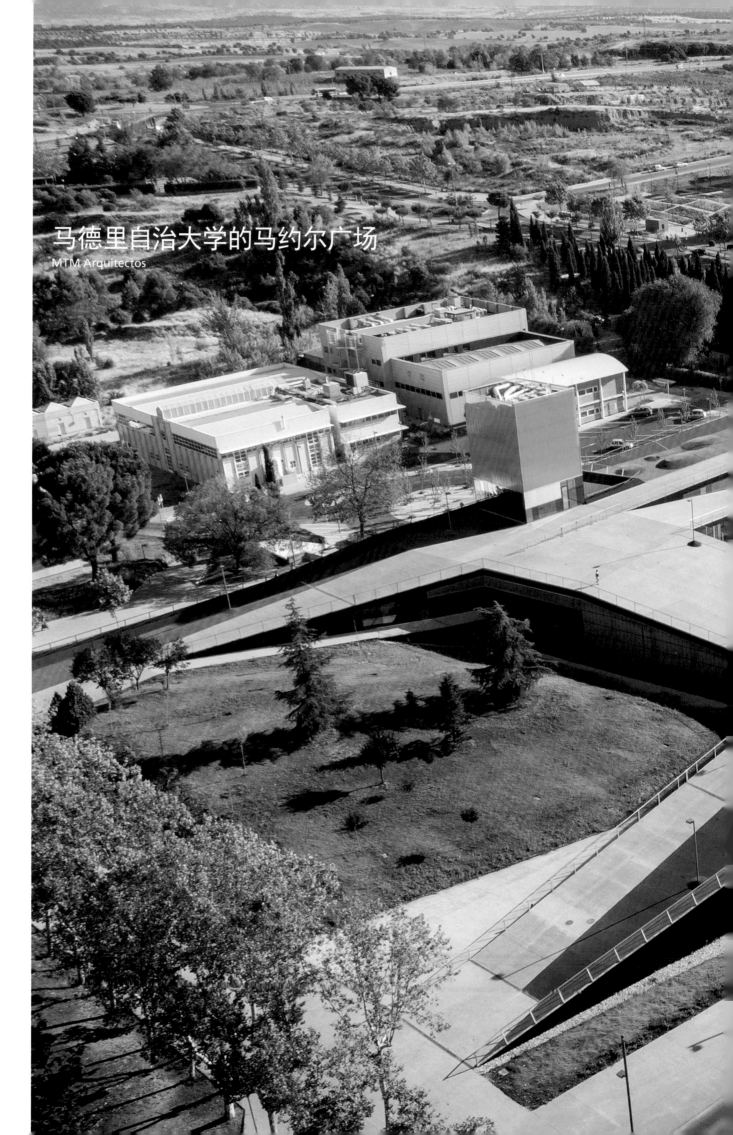

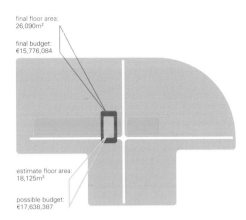

final floor area: 26,090m²
final budget: €15,776,084
estimate floor area: 18,125m²
possible budget: €17,638,387

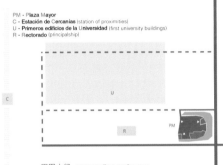

PM - **Plaza M**ayor
C - **Estación de C**ercanías (station of proximities)
U - **Primeros edificios de la U**niversidad (first university buildings)
R - **R**ectorado (principalship)

周围小径 surrounding pathways

连续的公共空间 continuous public spaces

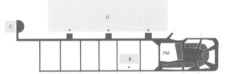

连续的人行道 (广场水平)　　▸ 建筑和服务区入口
continuous pedestrian　　　　　校园入口
(square level)
　　　　　　　　　　　　　　　access to buildings and services
　　　　　　　　　　　　　　　access to the campus

连续的人行道 (平台水平)　　▸ 校园入口
continuous pedestrian　　　　　建筑和服务区入口
(deck level)
　　　　　　　　　　　　　　　access to the campus
　　　　　　　　　　　　　　　access to buildings and services

 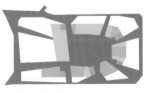

● 马德里自治大学室内服务区　　● 室外服务区
　internal services of UAM　　　　external services

● 垂直交流区　　　　　　　　　● 公共空间
　vertical communication　　　　common spaces

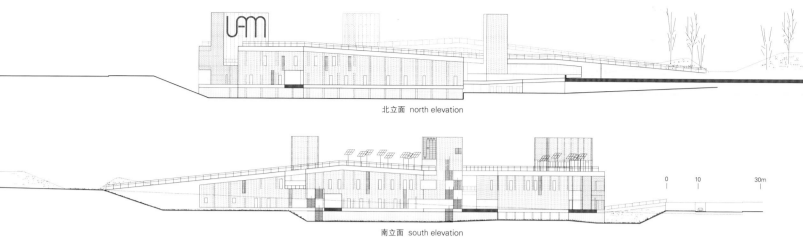

北立面 north elevation

南立面 south elevation

东立面 east elevation

修建的这座马约尔广场所在的城市有一个特殊的绰号——大学之城：马德里自治大学。广场的用户为年轻人或成年人、学生和老师，他们在这里进行着知识的传递。

该大学围绕着一个集会和交通中心空间拔地而起，且以绿色林荫大道为特色，这条大道已成为该大学的中央结构。与之毗邻的是科技大学、学区办公楼以及将人们与城市中心——太阳门广场连接的最近的换乘站。

在林荫大道的远端，马约尔广场围住了主要空间，即大学的生活空间。

鉴于林荫大道的横向封闭性以及与校园其他场地的自然连续性，马约尔广场必然成为该大学的新型服务核心。并且同时，它应表达其象征特性和代表性意义，作为校园结构的一种参照。

仅通过规模，密度和流畅性的融合使一处空间能够容纳现代人们性格的多样性。建筑的设计目的在于建造一处待定空间，灵活并且能够吸引大学用户将其用作预规划和非计划用途。

而其规模不允许限定具有清晰边缘定义的公共空间。

密度是定义其边界的方法：一个立方体结构形成了入口，并定义了其位置。

流畅性允许没有预定目的的自由和灵活的移动。

建筑师希望设计的环境没有障碍，让人们可以尽情地使用和享受，不存在开放和关闭时间，并且实现公共和私人空间的重合；人文环境通常受到所有权和隐私权的归类和限制，而这一方面首要的一点在于他们必须要以建筑师的角度开展工作。

人们穿过七座入口桥便可进入马约尔广场的内部。一进入内部，它便具有了城市标志：连续的、独特的，并且为庆祝活动提供能源服务。上部是林荫大道的阳台，两个不规则边缘的叠加平台是为了防止落叶飘落在地上。它既是一个广场又是一个拱廊，虽形式简单，却形成了一处广受到好评的空间。

屋顶是长度为800m的散步走廊的一部分，连接着火车站。它既是一个边缘，也是一扇凸窗，高出周围地面12m之多。从几何角度看，它也是一个截点，一个长达85m马蹄铁结构，传递着抵达感和吸引力。从此处可以望见锯齿状起伏的山脊，它如此之近，但有时却隐藏于校园所有公共空间的背后。

建筑周边的环境造就了现有的天台：照亮天台的一些灯柱与太阳能时钟同步，一些发光二极管发出微弱的白光，照亮了场地边缘和人工草坪、设计有不同颜色的橡胶辊和乙丙橡胶、光伏灯柱和饮品店。

项目使用了相同的材料，很自然，但在新学期开始也就是夏天结束的时候才能投入使用。

项目名称：Plaza Major Services Building
地点：Campus de Cantoblanco, Autonomaous University of Madrid, Spain
建筑师：Javier Fresneda, Javier Sanjuán
施工单位：Ferovial
结构工程师：IDEEE SL.
安装单位：Grupo JG.
外墙技师：Estrumaher
建筑成本估算师：Alberto Palencia, José Antonio Alonso
质量控制：Bureau Veritas
用地面积：17,662m²
总建筑面积：9,174m²
有效楼层面积：26,090m²
造价：urbanization_EUR 108/m², building_EUR 676/m²
竞标时间：2007
施工时间：2009~2012
摄影师：
©Ferrovial(courtesy of the architect) - p.16~17, p.18~19, p.20~21
©Luis Asin - p.22, p.29 top
©Roland Halbe - p.23, p.26~27, p.28, p.29 bottom

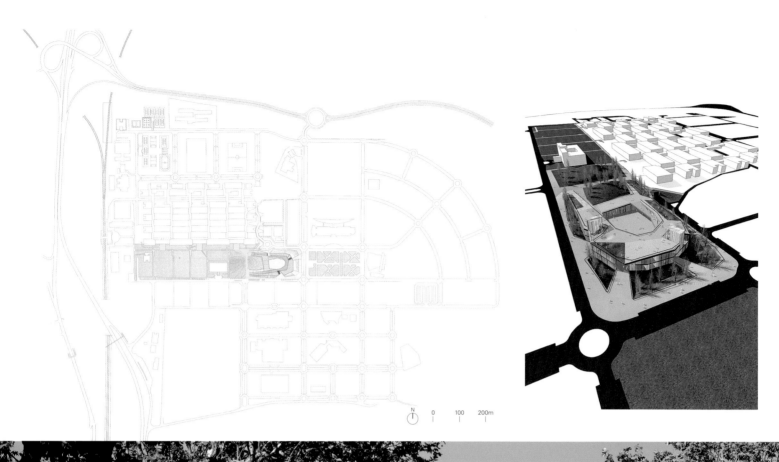

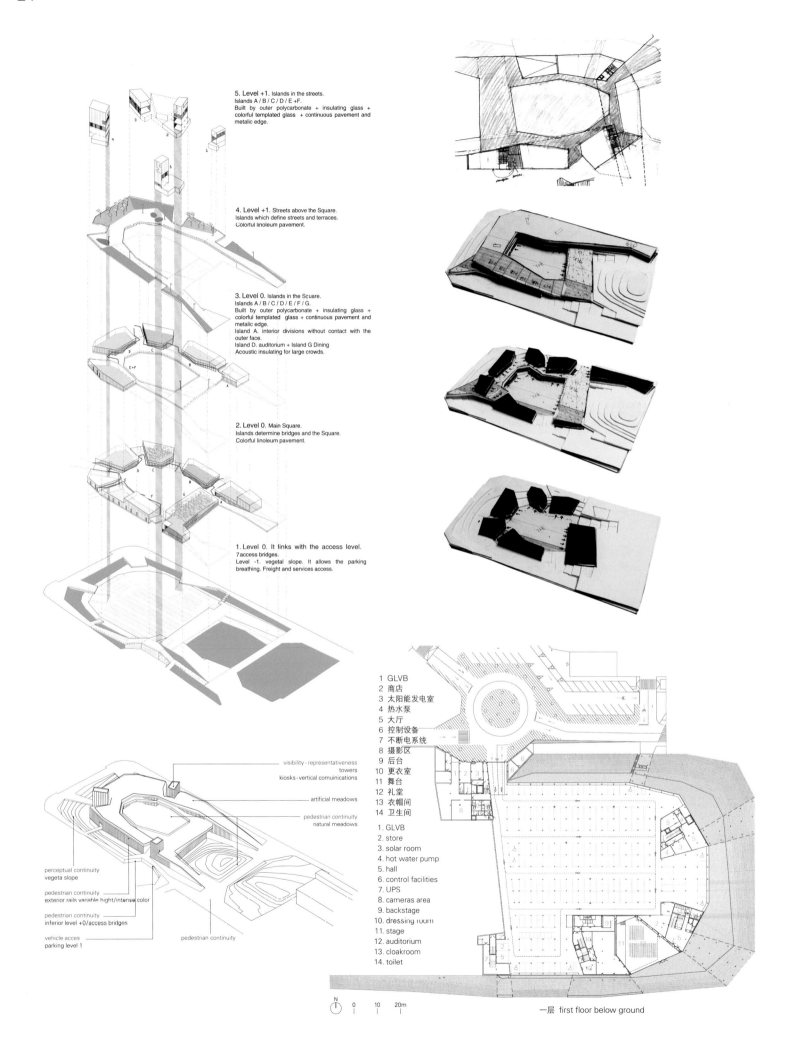

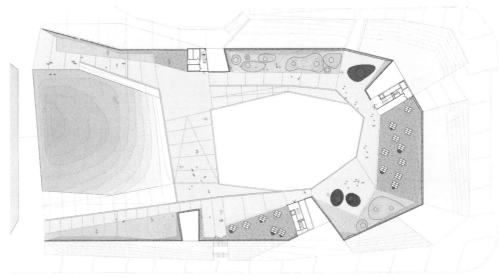

屋顶 roof

1 私人餐厅	11 洗衣房		
2 烹饪室	12 机房		
3 存储区	13 咖啡店		
4 设备区	14 旅行社		
5 公共入口	15 书店		
6 卫生间	16 复印室		
7 厨房	17 商店		
8 卸货区	18 工作室		
9 公共餐厅	19 办公室		
10 通知区	20 顾客服务区		

1. private dining room 11. laundry
2. cooking 12. computing area
3. storage area 13. coffee shop
4. facilities 14. travel agency
5. public access 15. book shop
6. toilet 16. reprographic area
7. kitchen 17. store
8. delivery area 18. work shop
9. public dining area 19. office
10. messaging area 20. customer service area

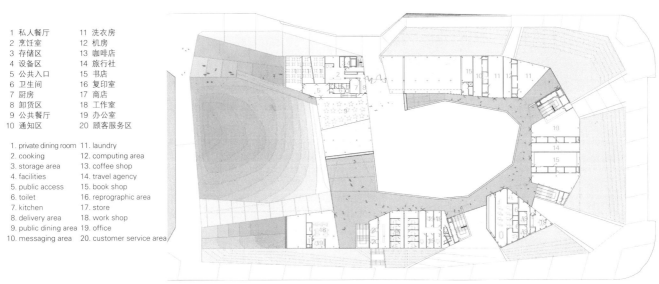

二层 second floor

1 储藏区 11 门廊
2 调味品区 12 银行支行
3 走廊 13 工作区
4 卫生间 14 办公室
5 自助区 15 药房
6 厨房 16 商店
7 餐厅 17 控制室
8 接待处 18 顾客服务区
9 展览区 19 会议室
10 公共广场 20 通道

1. storage area 11. porch
2. condiments area 12. bank branch
3. corridor 13. work space
4. toilet 14. office
5. self-service area 15. pharmacy
6. kitchen 16. store
7. dining room 17. control room
8. reception 18. customer service area
9. exhibition area 19. meeting room
10. public square 20. access

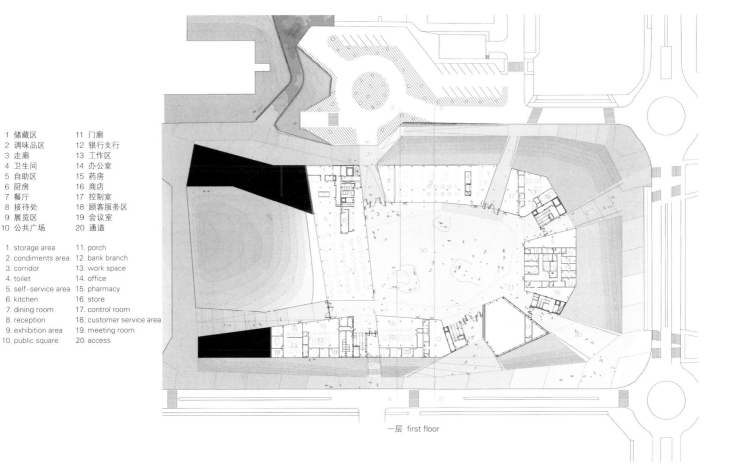

一层 first floor

Plaza Major of UAM

The city where the architects build this Plaza Mayor has got a specific surname, a University City: Autonomous University of Madrid. Users are youngers and adults, students and teachers, with their knowledge transmission.

The University has grown around a meeting and circulation central space, characterized by a green boulevard that has become the central structure of the University. Next to that are the Science University, the Rectorate and the recent commuter station which connects people with the City Center: Puerta del Sol.

At the far end of boulevard, the Plaza Mayor closes principal space, Living of University.

The Plaza Mayor must also be the new service core of the University, as the lateral closing of the boulevard and the natural continuity with the rest of the Campus. And, while, it should express its symbolic and representative nature, serving as a reference into the structure of the Campus.

Only by the means of scale, the density and the fluency can merge a space able to host the character variety of the contemporary individual. The target is to build an undetermined space, flexible and able to attract the university user both for the preprogrammed and non-planned use.

The scale doesn't permit to limit a public space, defined by legible edges.

The density is the mean to define the borders: the solids build and define the access.

The fluency allows a free and flexible movement, without a predetermined destination.

The architects like to work without barriers, with invitation to use and enjoy, without opening and closing hours, superposing the public space and the private space; The human surrounding are usually classified and regulated for the barriers of property and privacy, and its first point in this aspect is that they have to work as architects.

7 access bridges bring them to the interior of Plaza Mayor. Once inside it's identified with the city: continuous, unique, assisted for energy services. Above them are the balconies of the boulevard which defoliate in two superposed platforms with non-coincident edges. It's both a plaza and an arcade, creating a simple form, a recognized and appraised space.

The roof is part of the 800 meters long promenade which connects people to the train station. It is an edge and an oriel, elevated more than 12 meters above the surrounding ground. It's also geometrically a cutoff, a horseshoe more than 85 meters long which transmits the sensation of arrival and appeal. From it, it is possible to see the nearby sierra, so near but sometimes hides from all the public space of the campus.

The Project's surroundings build now the rooftop: some crosiers illuminate it synchronized to the solar clock, some leds with mild white lights show people the edge, artificial lawn, rolls of rubber and EPDM designed with different colors, photovoltaic posts and drink kiosks.

It is built with the same materials, natural, but can only invite to its use till the end of the summer, just the beginning of the new semester. MTM Arquitectos

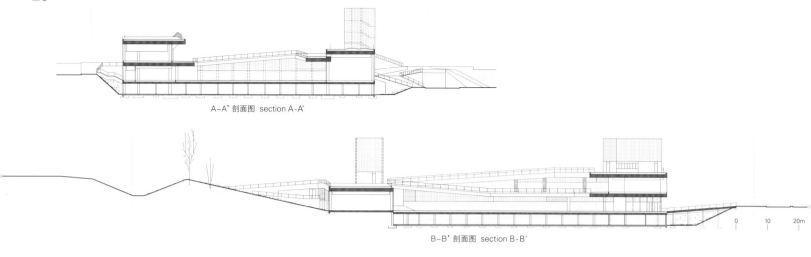

A-A' 剖面图 section A-A'

B-B' 剖面图 section B-B'

1. exterior enclosures

A) Heavy enclosures
HC Retaining wall. Reinforced concrete.

B) Light enclosures
Supporting substructure:
_ Perimeter profile made of anonized aluminium. Extruded aluminium which conformed tubular profiles RPT. Finishing anonized matt silver film (15 microns). The water/air thightness is obtained by EPDM joints.
_ Galvanized steel preframes #40.20.2
_ Galvanized steel substructure #120.60.4

M1 40mm / M2 25mm
Translucent wall made of alveolar polycarbonate panels with UV protection. Diagonal structure conformed by 6 walls Athermal. 40mm / 25mm.

V1 : insulating glass with low emissivity and solar control. Treatment layer on the inner side 10/1610 (interior colorless + holographic glass). Received on the both horizontal sides.

V1A: insulating glass with low emissivity and solar control. Treatment layer on the inner side. PVC sheet on the outer side 8+8/16/10 (interior colorless + holographic glass). Received on the both horizontal sides.

V2 : insulating and sheet glass with low emissivity and solar control. Treatment layer on the inner side 6/16/4+4 (interior colorless + holographic glass). Framed on 4 sides.

V3 : insulating and sheet glass with low emissivity and solar control. Treatment layer on the inner side 6/16/4+4 (interior holographic glass). Framed on 4 sides.

V4 : sheet glass 10+10 mm. (interior colorless + holographic glass). Framed on 2 sides.

V5 : sheet glass 5+5mm. (translucent). Framed on 2 sides.

2. interior enclosures

H25/30 Reinforced concrete.
F21 15 cm. Partition wall made of hollow bricks.
F2 27 cm. Partition wall made of hollow bricks.
Partition walls, dry assembled.
Galvanized steel profile, Ø=0,6mm. Pladur plasterboards.

A: interior isolation / LM: mineral wool. / w: exterior panels / wa: water-resistant / f: exterior fire-resistant panels / c_v: compound partition wall, variable thickness

A) Multiple *Pladurmetal* _ composition
T122   122/600(70)
T122-w   106/600(70)
TA122   122/600(70)LM
TA122-w   122/600(70)LM
T142   142/600(90)
T142-w   142/600(90)
TA142   142/600(90)LM
TA412   142/600(90)LM
TA142-f   150/600(90)LMf

B) Multiple *Pladurmetal* conformed by double structure
TcAv   variable/600(46 _chamber_46)LM

C) Self-supporting *Pladurmetal* plasterboard
Tr47-w   47/600(34)
Tr59   59/600(46)
Tr59-w   59/600(46)
Tr59   59/600(46)LM

3. finishing/coating

E: render with water-repellent cement mortar
G: finished in white plaster
P1: plastic plain matt paint
P2: Sealing (Stoneglaze VSC)
Z: galvanized plate in hot state, t=1,5mm
I: stainless steel plate, t=1,5mm
Osb: Oriented chipboard (OSB)
A: stoneware tiles 100x100

4. roof

Q1) Flat and non-accessible roof. Inverted roof finished with gravel.
_ 5cm. of pebbles 16/32 mm.
_ Thermal insulation 50mm. Extruded polystyrene panel (density 35Kg/m3) and geotextile separator sheet 150gr/m2.
_ Double Water-resistant sheet PA-8. Modified bitumen LBM 3+3.
_ 2cm. cement mortar for regularisation. 1%-5% slope above concrete floor.

Q2) Flat and accessible roof.
_ 2cm. of cement mortar.
_ Thermal insulation 50mm. Extruded polyesterene (density 35 kg/m3) and geotextile separator sheet 150 gr/m2
_ Double Water-resistant sheet PA-8. Modified bitumen LMB 3+3.
_ 2 cm. cement mortar for regulation. 1%-5% slope above concrete floor.
Qa2: continuous concrete pavement HA-25-B. Treated, t=7cm.
Q2b: syntetic monofilament grass, h=6mm.
_ Q2c: recycled rubber pavement. Granulated: EPDM. Coating: PUR.
_ Q2d: continuos in situ projection EPDM. Granule mixed with polyurethane resins PUR.

5. false ceilings

FT1: continuous and suspended plasterboard, Pladur Tec T-47/600/12,5
FT2: continuous plasterboard with circular drillings (R). Fibreglass cover film.
FT3: suspended by acoustic cylinders. Fiberglass wool. Coating made by fire-protected textile.

6. pavements

SO: continuous surface quartz treatment. Monolythic finishing. Hormipul.
S1: paint with Epoxy resins.
S2: continuous mortar with hight conditions Epoxi resins.
S3: continuous mortar with hight conditions Uretano resins. Thermal shock resistant. Non-skid treatment.
S4: Linoleum with polyurethane finishing. Eco-system, PUR. Amstrong DLW mod. Uni Walton Pur Eco System.
S5: prefabricated stair steps. Concrete HA-25-B. Treated, t= 35mm.
S6: continuous in situ concrete HA-25_b. Treated, t=7cm.
S7: fiber coconut doormat, t=24mm.

7. increase

R2: one-way floor with double self-supporting joist and hollow bricks 20-5cm. above brick partition walls (½ foot).
R3.60: 10cm. concrete with distributing mesh #6/15 cm. Polished finishing. Continuous filling EPS, IV, t=60cm.
R3.20: 10cm. concrete with distributing mesh #6/15 cm. Polished finising. Cellular concrete filling, t=20cm.

8. slabs

Fj0: concrete slab H30, t=2cm. Distributing mesh #8/20cm. Water-resistant polyethylene G400. Limestone layer 40/80, t=15cm.
Fj1: Suspended first slab conformed by air chamber, non-recoverable formwork and limestone layer, t=35cm.
Compression layer HA30, t=5 cm. with mesh #6/20cm. Air chambe, t=20cm.
conformed by prefabricated hollw bricks
PP-PET, t=2cm. Caviti. Concrete floor HA30, t=10cm. with mesh, #8/20cm. Draining layer.
Limestone layer 40/80, t=15cm.
Fj2: reinforced concrete slab, t=30cm.
Fj3: post-tensioning concrete slab HA30 with active steel Y 1860, t=35cm.
Fj4: reinforced concrete slab, t=25cm.
Fj5: collaborate slab, t=156mm. Self-supporting steel sheet, t=1mm.
Compression concrete layer HA30, t=8cm. with mesh, #5/15cm.

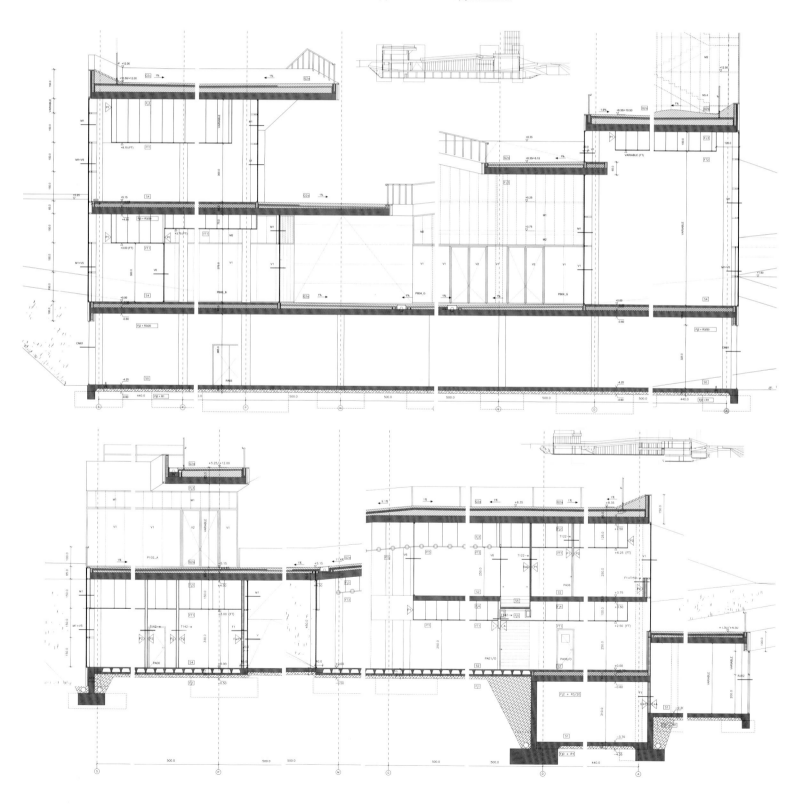

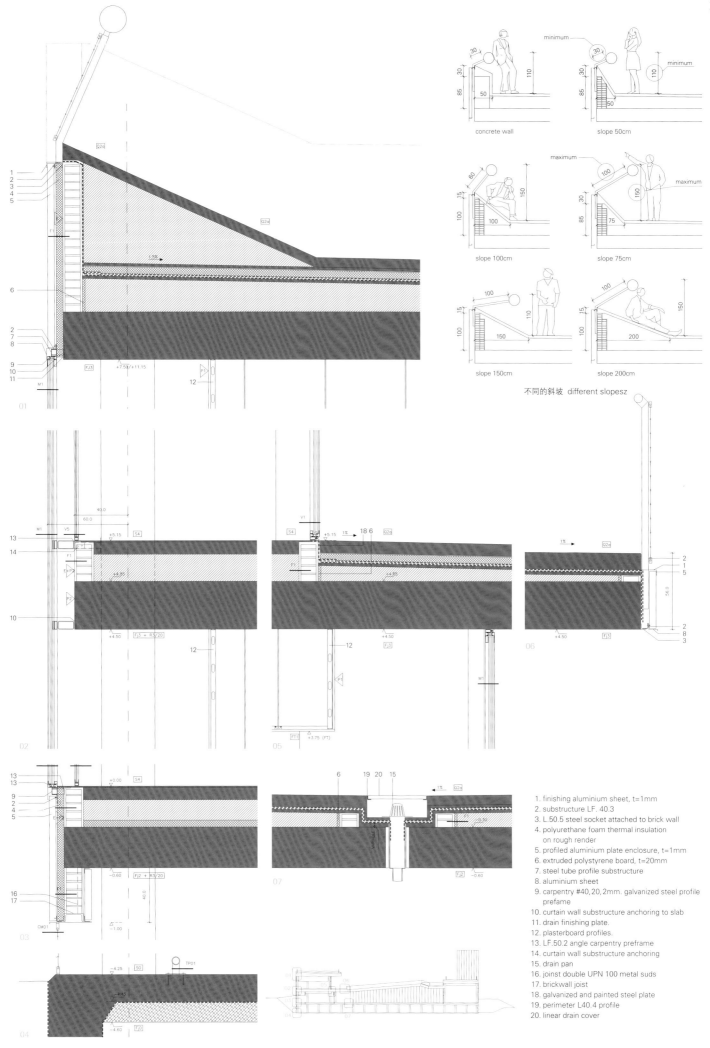

体验与感受

Experience and Sensing

艺术画廊和博物馆建筑在场地中发挥着重要的作用。作为社区的扩建物,文化建筑不仅为现有社区提供新型建筑设计和创意、社会和经济投入,来吸引远近的参观者,同时也与单个参观者形成一种密切的关系。空间中的每一瞥、每一次触碰以及每一个动作都记录着与感官层面的连接。

我们所穿梭的建筑和空间,直接或间接地影响着我们看待事物的方式、感受以及感知周围世界的方式。当我们参观举行艺术展览的建筑,在那里欣赏和体验艺术展览或表演时,会进行直接的感官体验,在那一时刻我们将与空间连接,产生一个特殊的场地。

作为参观者,由于我们在现实世界中存在着,在互动和移动时,与建筑的场地、规划、主题、材料以及感官品质形成回应,最终成为建筑组件的一部分。

在本文中,我们将从身体、参观者、住户或过路者的角度探讨和研究一些具有特色的美术馆和剧院建筑。主要的焦点将放在实际体验的理念、与建筑的互动以及建筑内部和远距离的处理方法上。

Art Gallery and Museum buildings have a significant role to play within the immediate site and location. As an addition to a community, the cultural building not only offers exciting new architectural design and creativeness, social and economic input to the existing community, attracting visitors from near and far, but also forms a close relationship to the individual visitor. Every glimpse, touch and movement through a space records and connects on a sensory level.

The buildings and spaces we move through, affect the way we see things, the way we feel and sense the world around us, directly or indirectly. Visiting buildings where art is on show, where we go to enjoy and experience art exhibitions or performances, immediately allows for sensory experiences, connecting us to the space at that very moment, creating a special place.

As visitors, we become part of the building components, through our very presence, as we interact and move through, responding to the site, programme, theme, materiality and sensory qualities of the building.

In this article, the featured Art Gallery and Theatre buildings will be studied and considered from the point of view of the body, the visitor, inhabitant or passer-by. Ideas of the experiential side of being in, interacting with and addressing buildings from within and from a distance will be the main focus.

cing

奥克兰美术馆 / Francis-Jones Morehen Thorp + Archimedia
柏林建筑图纸博物馆 / Speech Tchoban & Kuznetsov
水田艺术博物馆 / Studio Sumo
让•克洛德•卡里埃剧院 / A+ Architecture
乌镇大剧院 / Artech Architects
艺术馆 / Future Architecture Thinking
树美术馆 / Daipu Architects
丰岛横尾馆 / Yuko Nagayama & Associates
普图伊演艺中心 / Enota

体验艺术画廊与剧院 / Heidi Saarinen

Auckland Art Gallery/Francis-Jones Morehen Thorp + Archimedia
Museum for Architecture Drawings in Berlin/Speech Tchoban & Kuznetsov
Mizuta Museum of Art / Studio Sumo
Jean-Claude Carrière Theater/A+ Architecture
Wuzhen Theater/Artech Architects
House of the Arts/Future Architecture Thinking
Tree Art Museum/Daipu Architects
Teshima Yokoo House/Yuko Nagayama & Associates
Ptuj Performance Center/Enota

Experiencing Art Gallery and Theater Buildings / Heidi Saarinen

---

"我进入一栋建筑，看见一个房间，而在几分之一秒之内，我便对它产生了一种感觉。"[1]

一栋用于展示艺术和进行演出的大楼是如何与穿梭其中的人们的身体产生互动的？互动到底是如何产生的？它是怎样通过其结构、规划或是氛围来进行移动并且影响我们的？

作为创意空间的参观者，进入到美术馆、博物馆和演出场地的行为构成了与这些建筑进行直接互动的起点。我们进入到一栋建筑的目的可能是要欣赏艺术或体验一场演出，然而我们可能无法马上熟悉剧院、美术馆或博物馆建筑本身，对于举行艺术展览的建筑来说，我们也没有完全准备好接受其施加于我们身上的反应和力量。

当我们穿行于建筑和其各种内部空间、视角和观察点时，我们直接或间接地成为大楼的一部分。这种互动作用从单纯的空间参观和移动，进一步深入到实在的身体接触点。我们穿行建筑时所遇到的材料、氛围和气味的暗示唤起我们内心的记忆空间，将记忆与空间以及发生或想象的事件连接到一起。

我们与大楼形成一种感官上的连接，它是我们个人空间与我们出于不同原因（展览、剧院庭院中的一次社会活动或夜场演出）而参观的建筑的公共领域之间的一种秘密连接。

如何将建筑和空间体验成功整合进设计当中？在入口处初遇相识，我们作为接受者该作何反应？望向大厅、望向楼梯间上方、望入建筑深处的第一眼会以什么样的方式影响着我们？怎样做才是对建筑富有想象力的或经验性的回应？

不论我们是否暂时居住在该建筑中，作为参观者、表演者或是每天从外部观看的当地居民，我们在公共艺术建筑中度过时光的方式，以及回应特定建筑的参与方式，都是通过我们自身来将建筑与景观相互连接起来。实现建筑、其所处社区与独立个人之间的感官连接，是使该建筑的规划和用途得以成功实现的累积因素。

靠近建筑，我们触碰门把手或栏杆柱，或落座于剧院礼堂，必然会产生一处个人空间。"身体会产生认知和记忆。建筑的意义产生于身体和感官的过往回应和反应。"[2]

由A+建筑事务所设计的位于法国蒙彼利埃的让•克洛德•卡里埃剧院向我们讲述了一个引人注目的视觉故事。该建筑的肌理、材料和循环使用，共同发挥作用，应用于其错综复杂的宝石形格状立面中，并融合于外层表面、内部式样和由此产生的影子之间。这种抽象化表现的灵感来源于花斑眼镜蛇的形象，建筑以红色立面为背景，足以吸引参观者进入其中。外部的设计均参考于室内、屏风以及细部的风格，

"I enter a building, I see a room, and in the fraction of a second – have a feeling about it".[1]

How does a building of art and performance interact with the body moving through it? How, indeed, does a building move and affect us through its structure, programme or ambience?
As a visitor to creative spaces; art galleries, museums and performance venues, the act of entering forms the initial point of immediate interaction with the building. We may visit a building for the purpose of viewing art or to experience a performance, whilst we might not be directly familiar with the theatre, gallery or museum building itself, nor might we be quite prepared for the reaction and power the building hosting the art that may have on us.
Directly or indirectly, we become part of the building as we maneuver through it and its myriad of interior spaces, viewpoints and glimpses. The interaction goes further than seeing and moving through the space, to the actual physical point of contact. Hints of materiality, ambience and scents we come across as we travel through the building trigger our inner memory space, connecting memory to place and events occurred or imagined.
We form a sensory connection to the building, a secret link between our personal space and that of the public realm within the building we visit for different reasons: an exhibit, a social moment in the theatre courtyard or an evening performance.
How is this architectural and spatial experience successfully integrated into the design? As the first impression meets us at the entrance, how do we, the recipients, respond? How does the first glimpse into the foyer, over the stairwell, into the depth of the building affect us? What is the imaginative or experiential response to the building?
The way we spend time in public arts buildings, and the way in which we participate responsively to the particular architecture, whether we inhabit the building momentarily, as visitors, performers or by daily observations from the outside as local residents,

奥克兰美术馆,一系列精美的树形天篷覆盖并且形成了入口前院、中庭以及美术馆区域
Auckland Art Gallery, a series of fine tree-like canopies that define and cover the entry forecourt, atrium and gallery areas

而这些细部包括一些沿小径设计的灯具配件,采用了裸视的照明设备,暗示不同的心情和过渡,为参观者诠释并且照亮空间。

位于中国浙江的乌镇大剧院采取了对比场景,并设计了不同规模的并蒂莲造型,其设计上的戏剧化特征堪比剧院上演的戏剧和举行的节庆活动。美丽的诗意场景位于平静的湖水间,传统天台之上的钢架混凝土结构宛如一条蛟龙,从水面上奔腾而出。该建筑的双重表演空间(一个不透明,而另一个透明)以光、影和映像挑逗着彼此和观众。小型空间可灵活使用,受到了多学科从业者的青睐。

入口和出口点围绕着建筑的四周而设计,并且观众既可以从内部也可以从外部的门厅空间观看演出,在不同的季节和光照条件下从不同的角度来体验表演。

与所有的特色建筑一样,Francis-Jones Morehen Thorp + Archimedia设计的奥克兰美术馆,也毫无例外地追求用户与社区的互动关系。建筑师通过对一系列复杂并带有独立用途的新旧空间的有形阐述,将其整合为一个整体,使之与文化遗产建立起某种连接。反映景观的细节体现了对立面的概念,并参照临近公园的自然结构,使参观者和活动与周围社区建立连接。

大章建筑事务所设计的树美术馆位于中国北京宋庄的主城区,其外形的曲线和流动性在场景中发挥了重要作用。起决定性的延伸通道,以及高度不一的指引至今仍使人为之惊叹。设计规划中外形起到与场地连接的作用,而通过使用外部水源,参观者可以在此驻足深思,而当夜晚场地被灯光照亮时,这一场景尤为使人印象深刻。

该建筑通过活动、展览和戏剧演出,也在路线之间通过从公共区域到走廊和楼梯这些更私密的中间空间的简单互动来连接空间与用户。这些功能空间之间的体验进一步增加了建筑对人们的吸引力、人们的好奇心以及偶然性。

由Sumo工作室设计的日本坂户的水田艺术博物馆,位于日本榆树和樱花树环绕之间(或称浮世绘博物馆,译作"虚浮世界的图画",即日本木刻版画),它在其周围环境中形成互补,并融入到不同季节的景观当中,形成一种显著的表现方式。在建筑内部(大学校园的一

connect us with the building through ourselves to the landscape. Creating a sensory link between the building, its immediate community and the individual body, all add a layer of responsibility to the success and communication of the building's programme and use.

Close up, touching the door handle or bannister or sitting in the theatre auditorium, we inevitably create a personal stance to place. *"The body knows and remembers. Architectural meaning derives from archaic responses and reactions remembered by the body and the senses"*.[2]

The Jean-Claude Carrière Theater in Montpellier, France, by A+ Architecture, shows a striking visual narrative. The fabric, materials and circulation of use, all perform together, between this intricate diamond shaped lattice facade, merging between outer skin and interior pattern and consequent shadow. This abstract reference to the Harlequin, set against the red facade, draws the visitor in. Glimpses of the outside are seen referenced in the interior, in screens and details such as light fittings, whilst along some pathways, the lighting remains bare, hinting at various moods and transitions, narrating and illuminating the space for the visitor.

In a contrasting setting and at a different scale, the Wuzhen Theater, in Zhejiang, China, locally identified as the twin lotus, is as dramatic as the plays and festivals it stages. In a beautifully poetic setting, surrounded by tranquil water, the steel framed, reinforced concrete structure rises above the traditional rooftops as a dragon from the water appears to float on. Its dual performance spaces—one solid opaque, one transparent, take turns to tease each other and the audience with light, shadows and reflection. The smaller space is designed for flexible use, appreciated by multi-disciplinary practitioners.

Entrances and exit points are located around the building and the audience can view performances from inside as well as from the outside foyer space, experiencing the performance from interesting viewpoints in different seasonal and lighting conditions.

Like all the featured buildings, the Auckland Art Gallery by Francis-Jones Morehen Thorp + Archimedia, is no exception to aspiring to interact with the user and its community. The architects have made connections to the heritage through the tangible narrative of the complex set of spaces, new and old, all with an individual use, yet coming together as one. Details mirroring the landscape play with notions of opposites and refer to natural structures in a nearby park, connecting visitors and events to the neighbouring community.

In the Tree Art Museum by Daipu Architects, with its prime urban location in Songzhuang, Beijing, China, the curves and fluidity of the forms have a significant role in the setting. Decisive, sweeping walkways and differentiated levels guide yet allow for surprise.

水田艺术博物馆作为一座位于入口处的建筑，为人们提供了信息，并且展示了校园生活

Mizuta Museum of Art acts as a gateway building, providing information and displays about campus life.

建筑图纸博物馆位于普伦茨劳堡区的文化园区场地，其建筑展示了建筑图纸错综复杂的细节

Museum for Architecture Drawings, located in the cultural quarters of Prenzlauer Berg, showing off the intricate detailing of drawings

部分)有通道和画廊展示空间，吸引参观者，并为他们提供讲解。

该建筑将环境与其构件因素考虑在内，采用了混凝土、庇护式外形以及蓄热设计。较亮的玻璃表面使阳光能够渗透到开放空间。相比之下，其他区域被刻意地做了放暗处理，这样参观者就可以以最纯粹的方式欣赏艺术品了。这些感官上的细节设计，为参观者的体验增添了微妙的情感阐述。

通道光线和直线性的处理形成混凝土斜坡，允许走廊式通道中有一定的光线进入和移动，人们也可以在外部透过切口看到建筑的肌理。建筑师认为这种"有生命的、通气的通道使版画空间的欣赏者处于一个浮世之中"。

位于日本丰岛的丰岛横尾馆具备建筑和艺术的双重身份，是将2D和3D规范结合为一个建筑理念的项目创始人横尾忠则的作品。

随着当地大部分人口日趋老龄化，建立与社区之间的直接联系成为客户对建筑师的重要要求。建筑与社区之间关系的发展通过活动、互动以及自己动手活动得以促进，真正地使社区与建筑的使用和循环之间的关系更加密切。

建筑的起始点是一系列的传统木屋，它们被重建以用于新用途。延伸并且增添有趣彩色"拼贴画"的细节，如建筑通道当中的玻璃和镜子，使内部空间与外部空间在此相遇并映射出展出的艺术品。对颜色的重点使用、艺术品的放置以及内部和外部失真的景观色调，通过玻璃和视觉角度来赞美生命和死亡的阶段；将"普通与非凡"并置在一起。这种带有戏剧性，又使人有些熟悉的惊喜在该建筑中随处可见。

由Speech Tchoban & Kuznetsov设计的具有表现特征的柏林建筑图纸博物馆，位于普伦茨劳堡区的文化园区场地。采用分层混凝土表面，其立面凸出于其他相关的传统街道立面，展示建筑图纸错综复杂的细节是该博物馆设计的最初设想。细节的展开从建筑外部到内部，甚至在扶手这样微妙的细节设计上都相互呼应。

设计小型图纸展览空间对建筑师来说是一种挑战，它们通常比较复杂，需要近距离的观看，因此需要设计一系列房间式的隔间，以方便

The play with form on the plan connects the site, through use of the water outside, allowing the visitor to take a moment to reflect, particularly effective at night when the site illuminates.
This building connects the space with its user, through the events, exhibitions and theatre productions – but also between the lines, through the simple flirting across the space from the public areas to the more intimate in-between spaces of corridors and stairways. These experiences between the functional spaces further add to the intriguing, curious and accidental features.
Set amidst Japanese elms and cherry trees, the Mizuta Museum of Art, Sakado, Japan, by Studio Sumo or Ukiyo-e museum (translates as "Pictures of the Floating World", Japanese woodcuts), sits complimentary within its immediate surrounding, blending into the different seasons of the landscape, forming a prominent statement. Within the interior, part of a university campus, there are walkways and gallery spaces showcasing, enticing and educating the visitor.
The building takes into account the environment and its elements, using concrete, sheltering from and retaining heat. Glass appears as a lighter surface allowing daylight to penetrate the open spaces. Other areas, in contrast, are deliberately dark, for the art to be seen in its purest form. These sensory details, add a delicate emotional narrative to the visitor experience.
Manipulation of light in the walkways and the linearity that make up the concrete ramps, allow a playful amount of light and movements into the corridor-like passages, also seen from the outside, through cuts into the fabric. The architects explain that this "animates and aerates the passages, placing the viewer in the space of the print, within the floating world".
Teshima Yokoo House, located in Teshima, Japan, has an overlapping identity between the architecture and the art of project founder Tadanori Yokoo, combining 2D and 3D disciplines as one idea for the building.
With a largely aging local population, a direct connection to the community was an important requirement from the client to the architects. The relationship between building and community is encouraged through activities, interactions and hands-on events, genuinely uniting the community closer to the use and circulation of the building.
The starting point was a series of traditional wooden houses, remodelled into new use. Extending, and adding interesting chromatic "collages" of detail such as glass and mirror to the route through the building, allow the inside and outside to meet and reflect the art on show. Strong use of colour, placement of art pieces and distortion of hues of the landscape inside and out, through the glass and view points celebrate the stages of life and death; juxtaposing the "ordinary with the extraordinary". This sense of playful surprise, yet familiarity, is evident throughout the building.

树美术馆,其外形的曲线和流动性在场景中发挥了重要作用
Tree Art Museum, the curves and fluidity of the forms have a significant role in the setting.

普图伊演艺中心的实体设计,其中性的色调使其本身没有与重建部分复苏的美感相争艳
a solid design of Ptuj Performance Center, sufficiently neutral so as not to compete with revived beauty of the restored

在合适的距离观看这些陈列品。这些隔间与那些更小的展览空间,或者"陈列柜"一起,意味着需要限定每次的参观者数量,来为参观体验提供隐私。而繁忙时间在一层大厅图书馆等待观看藏品无疑是一种舒适的体验。

竞标获得者Enota设计的普图伊演艺中心位于斯洛文尼亚的普图伊,它有着丰富的建筑历史,应用也十分广泛,可以追溯至罗马、巴洛克和哥特的影响和居住形态。

从这一古代背景出发,通过谨慎的考虑,建筑师引入了当代扩建物,在建筑的内部与外部的肌理上产生一种与过去痕迹明显一致的特性。在挑战中,该项目成为一个复杂和持续的修复项目。建造计划内允许开展未来和正在进行的保护和修复工作,以及潜在的进一步考古发现。

参观者也确实可以感觉自己成为漫长的初露端倪的进程以及建筑和历史场地持续开发项目中的一部分。

对于Future Architecture Thinking设计的位于葡萄牙米兰达科弗的艺术馆来说,建筑师将其规划为一栋多功能艺术建筑,"人们相聚于庆祝的地方,在那里产生文化和艺术,它是一处促进和鼓励创造性活动的空间,也因此提高全体人民的生活质量"。

该建筑大型的红色形态对景观产生了直接的影响;虽起连接作用,但却似乎与主要内部空间,即剧院舞台区域、门厅和自助餐厅之间的一致性产生了一种冲突。各种高度的交通流线和大型体量,产生了进一步建造的机会。建筑师说道,"该项目的构思理念是设计通用的空间,技术上适用于不同类型活动的开展,以达到为全体人民服务的目的"。

作为参观者,建筑空间的体验可以呈现任何形式,从空想到文化交流的严肃目标。

"建筑的永恒任务乃是创造能被体验的、存在的隐喻,这种隐喻支持并且使人在世界上的存在性结构化。"3

建筑延伸人们与自然和景观的连接,而当我们的身体在这些经过构思、设计和开发的建筑地形当中时,我们需要的是一处能够实现多

A statement feature in Berlin, the Museum for Architecture Drawings, by Speech Tchoban & Kuznetsov is positioned into the site in the cultural quarters of Prenzlauer Berg. With its tiered concrete surface, the facade protrudes out of the otherwise relatively traditional street elevation, showing off the intricate detailing of architectural drawings that was the initial idea for the museum. The detailing carries through the building's exterior into the interior and is echoed as subtle details also in the handrails.

The architects had a challenge to accommodate the smaller drawings, often intricate, necessitating close up viewing, into a series of compartment like rooms suitable for viewing the exhibits from a comfortable distance. This, together with the smaller exhibition spaces, or "cabinets", means that visitor number is limited at any one time, adding intimacy to the experience. Waiting to view the collections at busy times is a comfortable experience, in the first floor lobby library.

The competition winning performance centre Ptuj Performance Center in Ptuj, Slovenia by Enota has a rich architectural history and use, tracing back to Romanesque, Baroque and Gothic influences and inhabitation.

From this ancient background, the architects introduced the contemporary additions with care and consideration, creating an identity alongside the clear traces of the past, in the fabric of the building throughout the interior and exterior. Not without challenges though, the project becomes a complex and continuing restoration project. Future and on-going conservation and restoration work and potential further archaeological finds are allowed within the building programme.

The visitor can certainly feel being part of an ethereal unfolding process and the continued development of the building and historical site.

House of the Arts by Future Architecture Thinking, located in Miranda do Corvo, Portugal, was planned by the architects as a multifunctional art building, *"celebrating the place where people meet, where culture and art happen, a space capable of promoting and stimulating creative activity, increasing the population's quality of life"*.

There is an immediate impact on the landscape – this large, red form, connecting but also appearing to gracefully collide in agreement between the main interior spaces: the theatre stage area, the foyer and the cafeteria. The circulation and sheer volume of the varying heights, evoke further opportunities. "The project was conceived by creating versatile spaces, technically suitable for different kinds of events, in order to serve all segments of the population", the architects state.

As a visitor, the experience can be anything from an architectural playing field of spaces inviting daydreaming to a serious destination for cultural exchange.

1. Peter Zumthor, *Atmospheres: Architectural Environments - Surrounding Objects*, Basel: Birkhauser, 2006, p 13.
2. Pallasmaa, J., *The Eyes of the Skin: Architecture and the Senses*, John Wiley & Sons, 2005, p 60.
3. Ibidem, p. 71.
4. Ibidem, p. 41.

感官体验的空间。⁴

我们会接近一栋在设计上全面考虑参观者感官体验和期望的建筑。如果我们容许建筑与我们产生交流并为我们指明方向,使建筑从平常无奇开始成为一种安抚和庇护所,并且带领我们开启建筑的想象和神奇之旅时,我们就会密切参与到这种交流之中,并且在那里我们期待建筑的功能能够得到满足,我们的好奇心可以被激起并且能有出人意料的体验。

我们可以在演艺中心和丰岛横尾馆这两个项目中找到该要素的概念。这是两座截然不同的建筑,一座用于举办演出,而另一座是用作美术馆。然而这两座建筑都采取有趣的建造方式,希望参观者强化体验并产生互动。

鉴于自己动手的工作室鼓励传统形式的手工和制作工艺,并且允许将其加入到最终设计方案当中,作为欢迎和信任地方性连接的一种直接证明,越来越多的老年人加入到其中。

而在乌镇大剧院的设计中,建筑语言更加正式,因为这里举行一系列大型活动、国际剧院节目,并且对材料和项目规划的要求极高。参观者可以从树美术馆的设计中发现相似之处,其连绵曲折的曲线和结实的材料形成创意活动的舞台背景。在让·克洛德·卡里埃剧院的设计中,娱乐和教育不仅仅发生在剧院的内部,当眼睛捕捉到地平线上宛如宝石般的网格立面时,旅程体验在一段距离之外就已经开始了,并且在周围场地中延续。

人们从水田艺术博物馆和建筑图纸博物馆可以看出形态、有触感的材料和整体的建筑设计,包括图案、规模和几何图形的布局。在奥克兰美术馆的设计中,材料的选择和形态,以及所处场地自然参照物的应用可以明确地看出其尊重过去的主题。

总之,材料、规模、比例实际上是实现一座建筑和其设计与我们交谈预期体验的途径,也是一座建筑产生空间与其用户–身体之间互动的方式。

*"The timeless task of architecture is to create embodied and lived existential metaphors that concretise and structure our being in the world."*³

Buildings extend connections to nature and landscape, and whilst our bodies' move through these considered, designed and developed architectural topographies, we are reminded of the need for spaces allowing multi sensory experiences.⁴

We get closer to a building where the design has fully considered the sensory side of the visitor experience and expectations. We become closer involved if we allow the building to communicate with us and show us the way, to the extent that the building starts to act as a comfort and refuge from the ordinary, and takes us on a journey into the imagination and wonder of architecture, where we can hope to be met by function, intrigue and surprise.

Elements of this notion can be found in the Performance Center and the Tashima Yokoo House. They are two very different buildings, one hosting performance and the other an art gallery, both inviting a playful approach from the visitor, expecting the visitor to add and interact to the experience.

With elderly people taking part in hands-on workshops where traditional aspects of craft and making techniques are encouraged and then allowed into the final design scheme as an immediate reference to the welcoming and trusting local connection.

Whilst in the Wuzhen Theater, the architectural language is more formal, inviting a range of major events, international theatre festivals and a hefty stance to materiality and programme. Similarly, the visitor is met at the Tree Art Museum, by its sweeping curves and solid materials setting the backdrop for the creative activities. The Jean-Claude Carrière Theater where enjoyment and education are not just taking place inside the theatre, but the experience starts some distance away and continues throughout the surrounding site, as the eye catches the jewel-like lattice facade in the horizon.

Characteristics of form, tactile materiality and overall architectural choreography are seen in the Mizuta Museum of Art and the Museum for Architecture Drawings, including arrangements of pattern, scale and geometry. In the Auckland Art Gallery the theme of honoring the past, is positively seen in the choice of material and form and in the use of natural references to the immediate site.

To conclude, the materials, scale and proportion are indeed the way that a building and its programming talks to us about what we are expected to experience, and how the building allows for the interaction between the space and its user, the body.

Heidi Saarinen

奥克兰美术馆

Francis-Jones Morehen Thorp + Archimedia

新建的奥克兰美术馆Toi o Tāmaki是一个设计类型广泛的公共项目,包括恢复和改造遗产建筑、扩建新建筑以及重新设计临近的阿尔伯特公园。

这座建筑从一个理念开始发展,这一理念与景观的有机自然形式紧密关联,使其与遗产建筑的建筑顺序和特色有着密切的联系。

新建筑的特色为一系列精致的树形天篷,以此形成并且覆盖入口前院、中庭以及美术馆区域。这些轻质的外形的设计灵感来源于附近Pōhutukawa树(新西兰圣诞树)的树冠以及其盘旋于石墙和露台(重新诠释了场地的自然地形)的形式。天篷的天花板是由精选的贝克杉树制成,切割成精确的几何形状,由细长的、顶端尖尖的杆子进行支撑。这些具有象征性的外形赋予了美术馆一个独特的特色,即它是被场地的自然景观赋予灵感的。

阶梯石台与盘旋的树冠之间形成了一种开放性和透明性,允许人们的视野穿透美术馆的交通流线和展览空间,进而进入阿尔伯特公园的绿色景观之间。以此,美术馆以邀请和加入的姿态,面向公园和附近的公共区域开放。

进入美术馆的入口次序遵循着从街道一侧的前院、宽敞的迎宾天篷区域的下方,穿过较低的门厅区域,进而通过一个大型楼梯,进入大型的、明亮的中庭的序列。中庭为所有的访客提供了中央方位以及展览空间。美术馆的交通流线从主中庭开始延伸,形成一个清晰的环状外形,通过南侧的小型中庭,将所有的美术展览空间连接起来。

项目创建多元化的展览空间和房间,有固定的和灵活移动的、正式的和非正式的、古老的和现代的、自然采光的与人工照明的、开放的和封闭的,以及高度不一的。

Auckland Art Gallery

The new Auckland Art Gallery Toi o Tāmaki is an extensive public project that includes the restoration and adaption of heritage buildings, a new building extension, and the redesign of adjacent areas of Albert Park.
The architecture has developed from a concept that relates as much to the organic natural forms of the landscape as it does to the architectural order and character of the heritage buildings.
The new building is characterized by a series of fine tree-like canopies that define and cover the entry forecourt, atrium and gal-

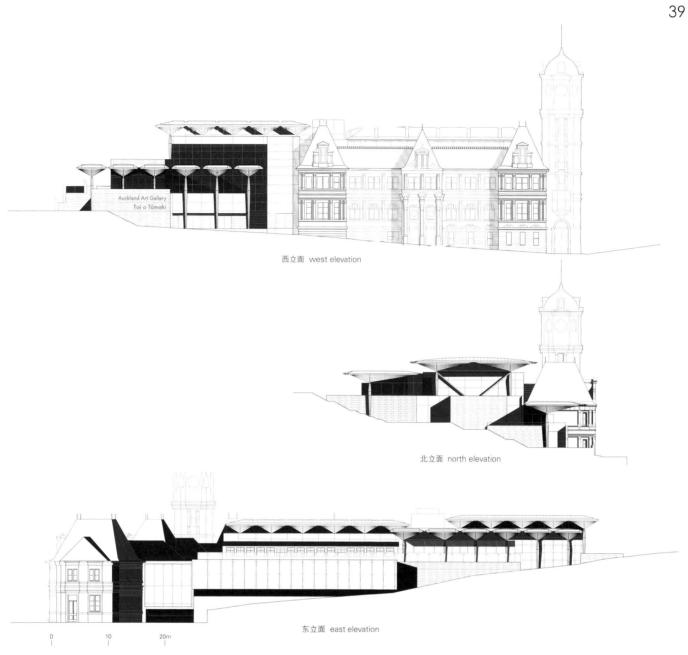

西立面 west elevation

北立面 north elevation

东立面 east elevation

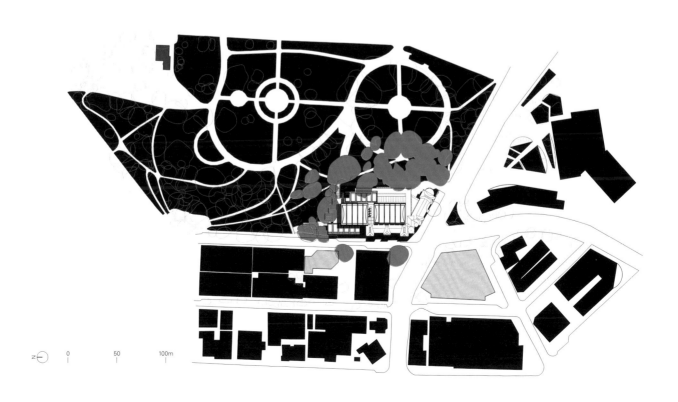

lery areas. These light, profiled forms are inspired by the adjacent canopy of Pōhutukawa trees and "hover" over the stone walls and terraces that reinterpret the natural topography of the site. The ceilings of the canopies are assembled from carefully selected Kauri, profiled into precise geometric patterns and supported on slender and tapering shafts. These emblematic forms give the gallery a unique identity that is inspired by the natural landscape of the site.

Between the stepped stone podium and hovering canopies,

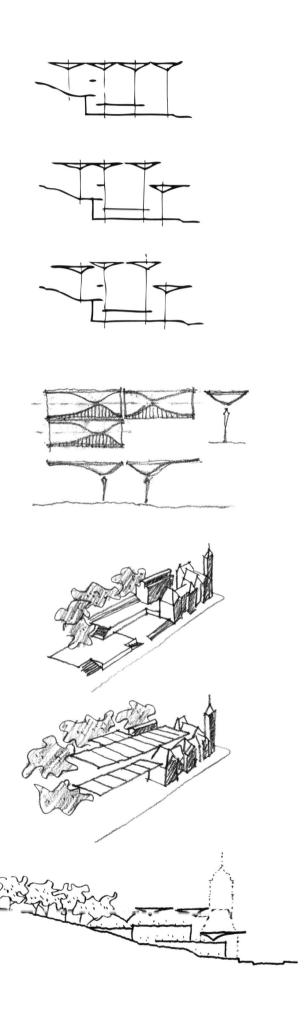

openness and transparency are created to allow views through, into and out of the gallery circulation and display spaces and into the green landscape of Albert Park. In this way the gallery opens to the park and adjoining public spaces in an inviting and engaging gesture of welcome.

The entry sequence into the gallery follows a progression from the street forecourt, under a generous and welcoming canopy, through into a lower foyer to emerge via a broad stair into the large, light-filled atrium. The atrium provides a central orientation and display space for all visitors. Gallery circulation extends from the main atrium in a clear series of loops interconnecting all gallery spaces via the smaller southern atrium that mediates the junction with the existing Wellesley Wing.

A diverse range of exhibition spaces and rooms are created, both fixed and flexible, formal and informal, heritage and contemporary, naturally lit and artificially lit, open and closed, high spaces and lower spaces.

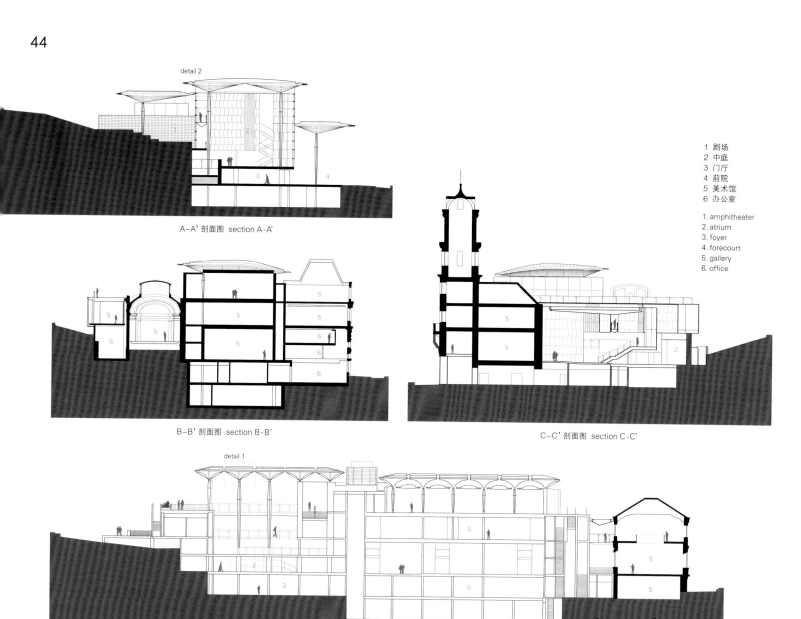

A-A' 剖面图 section A-A'

B-B' 剖面图 section B-B'

C-C' 剖面图 section C-C'

D-D' 剖面图 section D-D'

1 剧场
2 中庭
3 门厅
4 前院
5 美术馆
6 办公室

1. amphitheater
2. atrium
3. foyer
4. forecourt
5. gallery
6. office

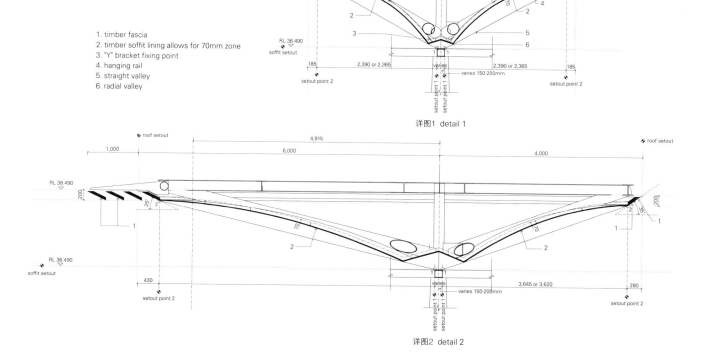

1. timber fascia
2. timber soffit lining allows for 70mm zone
3. "Y" bracket fixing point
4. hanging rail
5. straight valley
6. radial valley

详图1 detail 1

详图2 detail 2

项目名称：Auckland Art Gallery Toi o Tāmaki
地点：Auckland, New Zealand
建筑师：FJMT + Archimedia
项目团队：FJMT_Richard Francis-Jones, Christine Kwong, Alexander Pienaar, Eric Lee, Martin Hallen, Brooke Matthews, Phillip Pham, Michelle Ho, Jeff Morehen, Richard Desgrand, Matthew Mar, Daniel Schagemann, Katty Huang, Zuzana Semelak, Matthew Todd/Archimedia_Lindsay Mackie, Neil Martin, Russell Pinel, David Pugh, Surya Fullerton, Hamish Cameron, Yogesh Dahya, James Raimon, Sakouna Traymany, Edwin Chen, Sen How Tan, Jaime Don, Shaun Wong, Damon Aspden
项目管理：Coffey Group
结构：Holmes Consulting Group
音响：Marshall Day Acoustics
景观：Melean Absolum Limited
照明：Steensen Varming
甲方：Auckland Regional Facilities
承包商：Hawkins Construction
用地面积：3,642m²
有效楼层面积：14,370m²
造价：NZD 121m
竣工时间：2011.9
摄影师：
©John Gollings (courtesy of the architect) - p.38, p.40~41, p.42~43, p.46
©Patrick Reynolds (courtesy of the architect) - p.45

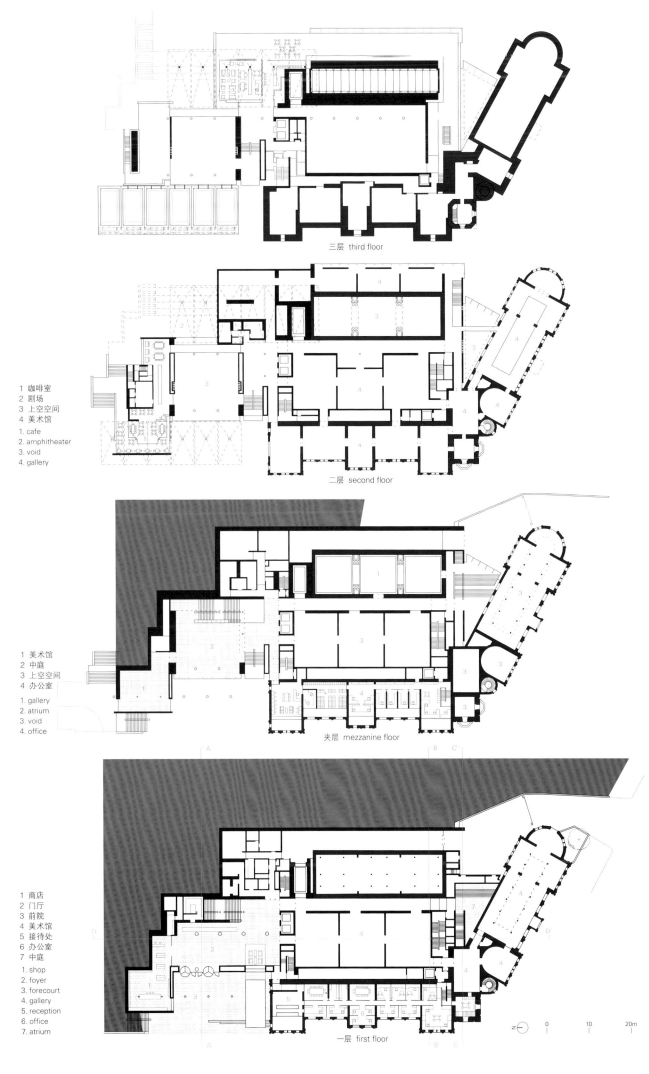

三层 third floor

1 咖啡室
2 剧场
3 上空空间
4 美术馆

1. cafe
2. amphitheater
3. void
4. gallery

二层 second floor

1 美术馆
2 中庭
3 上空空间
4 办公室

1. gallery
2. atrium
3. void
4. office

夹层 mezzanine floor

1 商店
2 门厅
3 前院
4 美术馆
5 接待处
6 办公室
7 中庭

1. shop
2. foyer
3. forecourt
4. gallery
5. reception
6. office
7. atrium

一层 first floor

柏林建筑图纸博物馆
Speech Tchoban & Kuznetsov

Art Gallery and Theater
艺术画廊与剧院

柏林建筑图纸博物馆项目意味着将2009年成立的Sergey Choban基金会的收藏品对外开放,以达到建筑图形的艺术推广以及不同机构(包括著名的伦敦约翰·索恩爵士博物馆或巴黎艺术学校)的临时展览目的。

为了建造博物馆,基金会在前Pfefferberg工厂区购买了一小块场地,这片区域是艺术的集聚地。这里还坐落着著名的艺术画廊ADEDES,一座现代艺术画廊和艺术工厂。而正在建造中的建筑图纸博物馆则成为普伦茨劳堡区新型文化中心(在柏林人中备受欢迎)发展的一个连续。

新博物馆建筑位于临近四层住宅楼的防火墙的一侧。在当前的发展条件下,这里的临近建筑和场所暗示了博物馆的非常规性空间规划布局。从设计角度来看,紧凑的体量上升至邻居的屋脊处,形成了插进建筑中的五个体块,并且相互制约。最上层的体块由玻璃制成,以屋檐的形式悬挂于整座建筑之上,其余四个较低的体块的立面则由混凝土制成,其表面绘制了浮雕式的建筑细部图,并且在每个楼层上加以重复,彼此重叠,以形成图纸状。这种艺术触感能够在博物馆的建筑形象中突出其功能和展览的内容。

Christinenstrasse街一侧的三层和四层的大型混凝土墙体的表面十分平坦,并且与大型玻璃涂料交相变换,以突出主建筑入口,以及拥有图纸立面的小型陈列室前侧的重建区域。一层设有一个入口大厅,即图书馆。两间小型陈列室位于上层,以展览图纸和文档。所有楼层都通过一座电梯和楼梯相互连通。

项目名称:Museum for Architectural Drawing in Berlin
地点:Christinenstrabe 18, Prenclauer-Berg, Berlin, Germany
建筑师:Sergei Tchoban, Sergey Kuznetsov
首席建筑师:Philipp Bauer, Ulrike Graefenhain, Ekaterina Fuks
项目团队:Dirk Kollendt, Nadezhda Fedorova, Frederik S. Scholz
合作商:nps tchoban voss GmbH & Co.
总建筑面积:279m² 有效楼层面积:498m²
设计时间:2009 施工时间:2011~2013
摄影师:
©Patricia Parinejad(courtesy of the architect)-p.53, p.55
©Roland Halbe(courtesy of the architect)-p.48, p.49, p.50, p.51, p.54

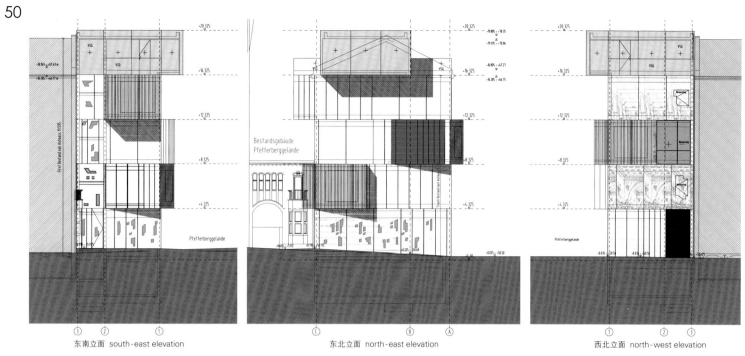

东南立面 south-east elevation　　东北立面 north-east elevation　　西北立面 north-west elevation

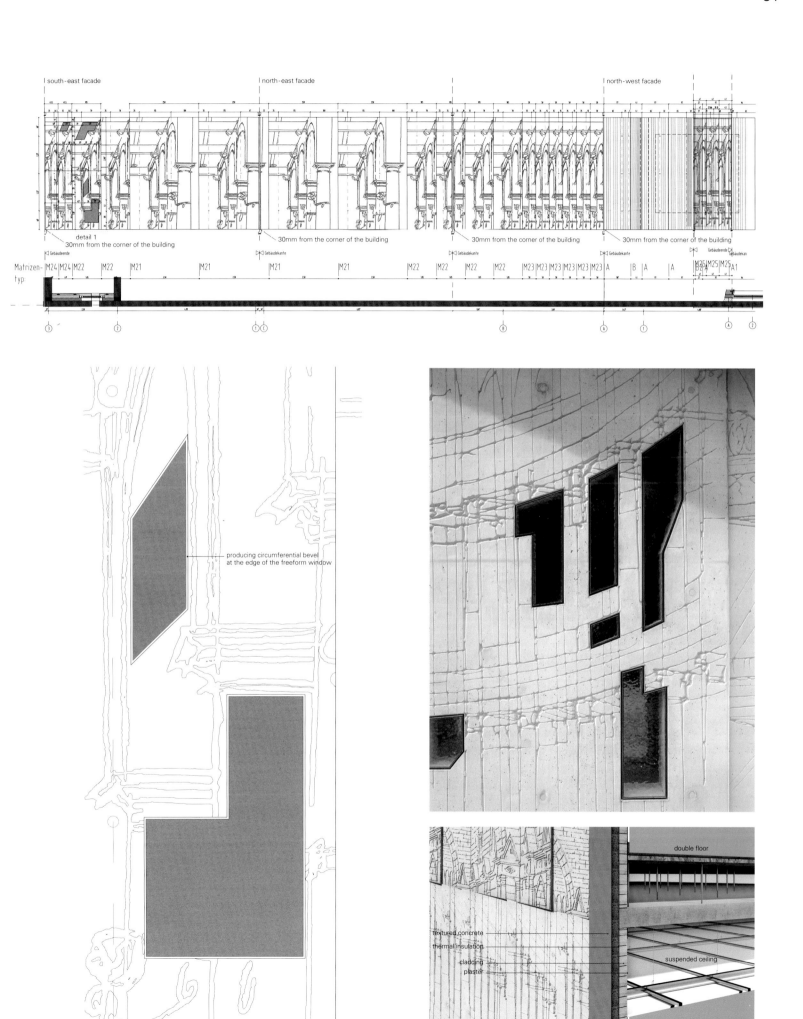

详图1_自由形式的窗户图像详图 detail 1_freeform window cutout detail

Museum for Architecture Drawings in Berlin

The Museum for Architectural Drawing is meant for placing and exposing the collections of Sergey Choban's Fund founded in 2009 for the purpose of architectural graphics art popularization as well as for interim exhibitions from different institutions including such famous as Sir John Soane's Museum in London or Art school in Paris.

For the construction of the Museum, the Foundation purchased a small lot on the territory of the former factory complex Pfefferberg, where the art-cluster is formed. Here are already located the famous architecture gallery AEDES, the modern art gallery and artists' workshops. The Architectural Drawing Museum that is being constructed became a logical continuation to the development of the new cultural center in a district Prenzlauer Berg that is very popular among Berlin residents.

The new Museum building flanks the firewall of the adjacent four-storey residential house. Such neighborhood and the location under the conditions of the current development implied the irregular space-planning arrangement of the Museum. The volume that is compact in terms of design rises up to the mark of the neighboring roof ridge, forming five blocks clearly cut in the building carcass and offset in relation to each other. The upper block, made of glass, hangs over the whole volume of the building in cantilever. The facades of the four lower blocks are made of concrete and its surfaces are covered with relief drawings with architectural motives, repeating on every level and overlapping each other as sheets of paper. This artistic touch is supposed to emphasize the function and contents of the exposition in the Museum architectural look. On the first and third floors from the side of Christinenstrasse, the flat surfaces of the massive concrete walls are alternate with large glass paintings accentuating the main building entrance and recreation area in front of one of the graphic cabinets. On the first floor there will be the entrance hall – library. Two cabinets for drawings exposition and archive are located on the upper floors. The levels are connected by an elevator and stairs.

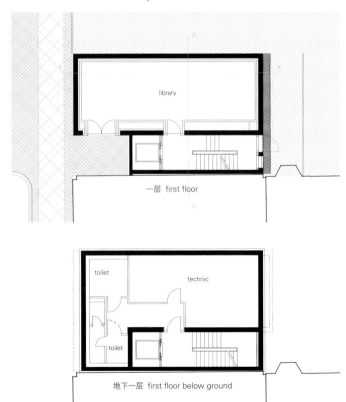

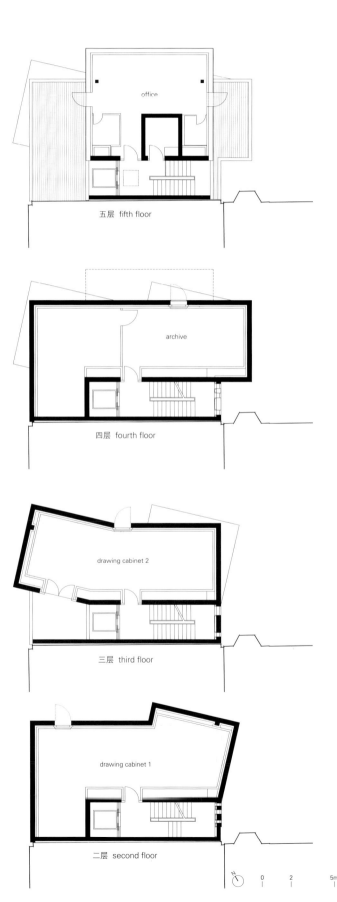

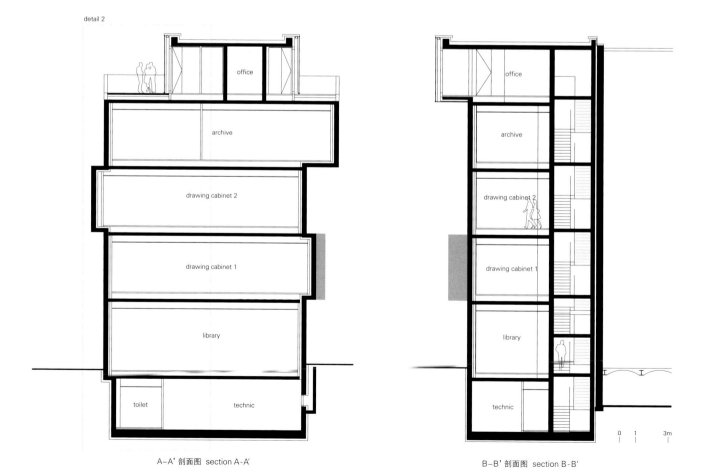

A-A' 剖面图 section A-A'

B-B' 剖面图 section B-B'

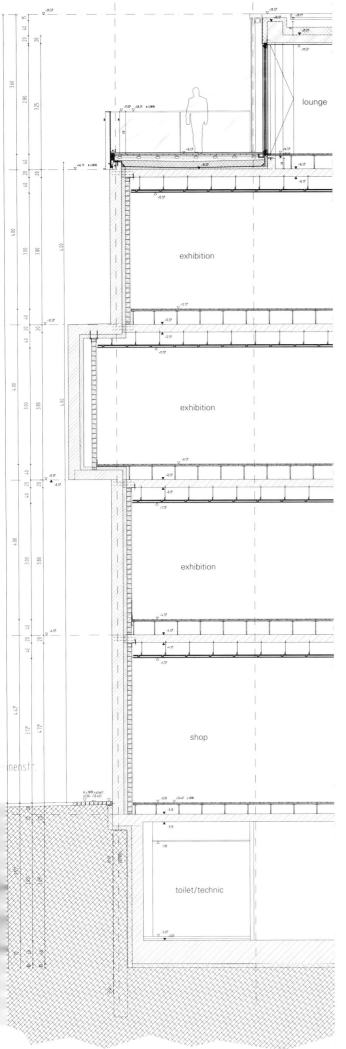

详图2 detail 2

水田艺术博物馆
Studio Sumo

水田艺术博物馆位于日本一所大学的校园内，是一栋面积为650m²的建筑，该博物馆的设计目的是展示一些浮世绘（日本木刻版画）的珍贵藏品，同时也可以展出一些当代的艺术作品以及学校和社区的美术作品。鉴于它们具有易碎的特性，这些版画需要一处高度控制和绝热的环境。而因为该建筑最接近校园入口，因此它同时也是一栋门户建筑，用于公告信息以及展示校园生活。紧凑但外露的场地保留了现存的17棵树木，并且建筑限高9m，所以最终形成了一座具备双重身份（博物馆和信息中心）的二层建筑，且每一个功能区都设有直接进入主校区的通道。

为使两个楼层都能够直通步行道，建筑场地的地下挖掘出半个楼层，一条坡道向上通往两个博物馆画廊，而另一条则向下通往校园信息中心。这些坡道的规模设计适中，可以负荷一定的重量，也适合用作公共入口，因此无需另外安装货梯。东入口的机械空间和西侧用作俯瞰风景的休息室与坡道空间一起，共同形成了一处外围环境缓冲区，避免阳光直射到画廊墙壁的外侧。

"浮世绘"译作"浮生世界的绘画"，这种版画通常描述著名演员、交际花或出行场景，意在使赏画者从其日常生活当中解脱出来。建筑师将这个概念转移成支撑信息空间上方画廊的现浇结构的漂浮构造。此外，立面的光槽设计使人联想到许多版画画作描绘雨的场景。

L形预制混凝土构件排列在建筑坡道上，形成了连续的垂直和水平遮蔽处。这些构件固定在现浇结构的屋顶之上，以及向上通往博物馆的斜坡或校园小径一侧的地面上。这52个构件都略有不同，由单个模具浇筑而成，在四个月的预制时间里每隔一天才能产生一块嵌板。每块嵌板约为1.2m宽，立起来约为8.5m高，长度为3.6m，并且其厚度不超过0.25m。

这些构件是由茨城县的一个工厂制造的，经卡车运输约161km到达东京北部的场地。与预制的嵌板不同，这些构件的两面都进行了浇筑，形成光滑的双面抛光裸露表面。长度不同的、宽度约为0.3m的缝隙沿着接缝线进行了封闭，某些缝隙从混凝土构件的垂直面上一直延续到水平面上。这便形成了光缝，使通道充满了生气与新鲜空气，让参观者处在"浮生世界"的画作空间内。

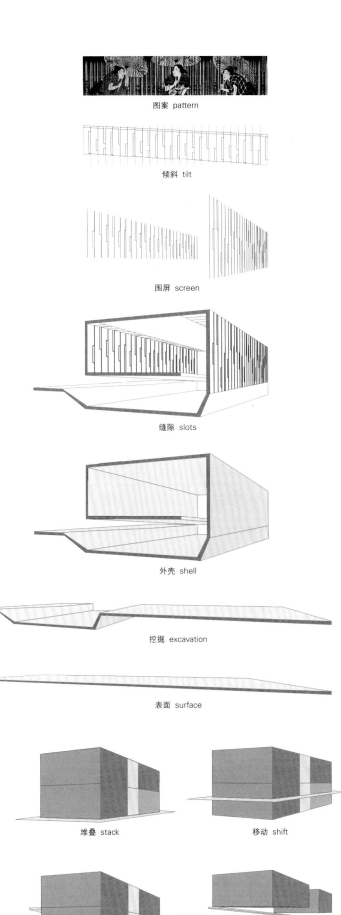

项目名称：Mizuta Museum of Art
地点：Sakado, Saitama-ken, Japan
建筑师：Sunil Bald, Yolande Daniels
项目团队：David Huang, Edward Yujoung Kim, Andrea Leung, Anees Assali
纪录建筑师：Obayashi Design Department, Obayashi Corporation
结构工程师：Obayashi Corporation
MEP工程师：Obayashi Corporation
承包商：Obayashi Corporation
甲方：Josai University
用地面积：995m² 总建筑面积：600m²
竣工时间：2011.12
摄影师：
©Daici Ano (courtesy of the architect) - p.56, p.58, p.59, p.60, p.61, p.63, p.64
©Koichi Tomura (courtesy of the architect) - p.62, p.65

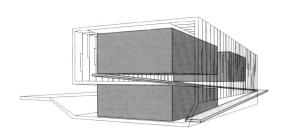

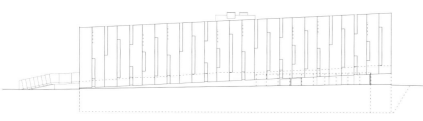

西北立面 north-west elevation

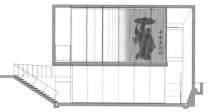

西南立面 south-west elevation

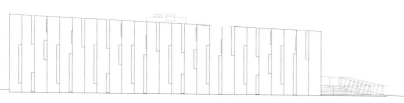

东南立面 south-east elevation

0　　5　　10m

东北立面 north-east elevation

Mizuta Museum of Art

Mizuta Museum of Art is a 7,000sf building in a university campus in Japan designed to show works from a valuable collection of Ukiyo-e (Japanese woodcuts), while also being able to accommodate contemporary works and artistic production from the school and community. The fragile nature of these prints requires a highly controlled and insulated environment. As the building closest to the campus entry, it also acts as a gateway building, providing information and displays about campus life. Finally, its compressed but exposed site with seventeen existing trees and nine meter height limit could only yield a two story building with a dual programmatic role, museum and orientation center, each requiring direct access to the main campus walk.

To give both floors equal access to the pedestrian route, the building is excavated a half level into the site, with one ramp leading up to the two museum galleries and another leading down to a campus information center. These ramps are dimensioned for loading as well as entry, thus eliminating the need for a freight elevator. In conjunction with the mechanical space at the east entry and an overlook lounge at the west end, the space of the ramp creates a perimeter environmental buffer that protects the exterior side of the gallery walls from direct sunlight.

"Ukiyo-e" is translated into "Pictures of the Floating World" as the prints often depicted famous actors, courtesans, or travel scenes meant to lift the viewer from the daily routine of his/her life. The architects were taken by this notion, which was architecturally

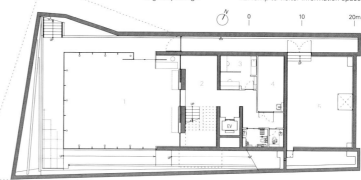

1. visitor information space
2. school history display
3. office
4. storage
5. mechanical room
6. gallery lounge
7. flexible gallery
8. entry foyer
9. reception
10. overlook
11. ukiyo-e gallery
12. ramp to visitor information space

地下一层 first floor below ground

north patio

campus walk

二层 second floor

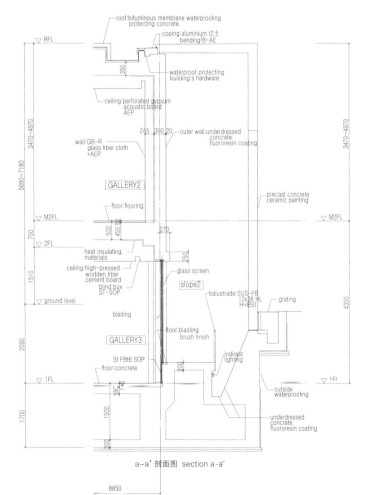

a-a' 剖面图 section a-a'

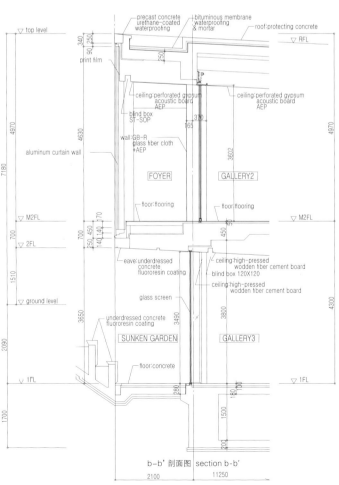

b-b' 剖面图 section b-b'

translated into the floating tectonics of the cast-in-place structure that cradle the galleries over the information spaces. Additionally, the slots of light created by the facade that recall the graphic method of depicting rain found in many of the prints.

L-shaped pre-cast concrete pieces line the building ramps and form a continuous vertical and horizontal shelter. These pieces clip onto the roof of the cast-in-place structure and to either the ramp leading up to the museum, or to the ground on the side of the campus walk. There are 52 pieces, and while each is somewhat different, all are cast from a single mold, one panel every other day for almost four months. Each is about four feet wide, up to 28 feet in along the vertical, up to 11 feet overhead along the horizontal, and less than 10" thick.

The pieces were fabricated in a factory in Ibaraki Prefecture and trucked about 100 miles to the site, just north of Tokyo. Unlike flat pre-cast panels, they were cast on their sides, resulting in two smooth finished surfaces that are exposed. One-foot wide slots of varying lengths were blocked out along the seam lines, some continuing for the vertical to horizontal section of the piece. This creates light slits that animate and aerate the passages, placing the viewer in the space of the print, within the "floating world."

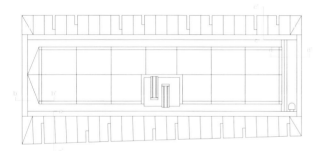

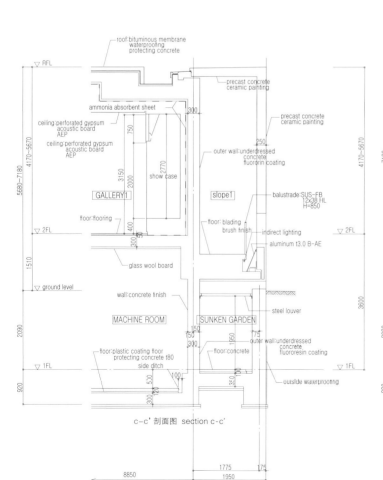

c-c' 剖面图 section c-c'

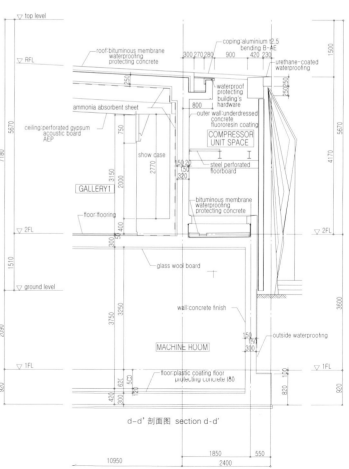

d-d' 剖面图 section d-d'

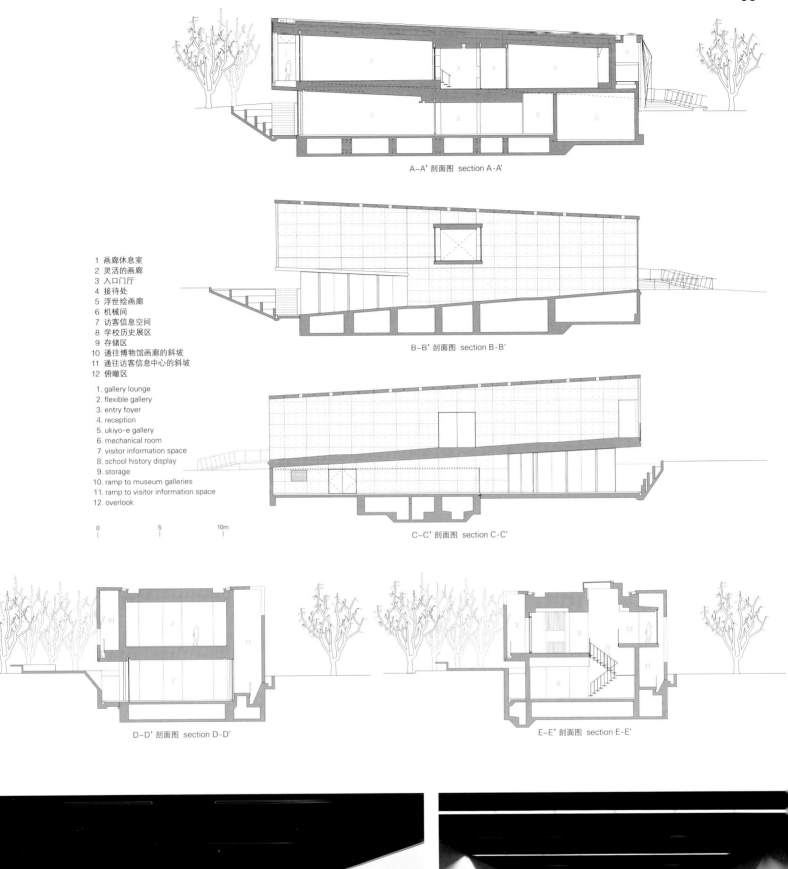

让·克洛德·卡里埃剧院
A+ Architecture

设计与选址

建筑师将该项目选址在Domaine de l'O（一个葡萄酒庄园，译者注）是令人惊叹的。它的松树林、它的历史以及它所覆盖的空间，都激发出了一个奇思妙想的理念。该剧院位于种满大树的大道与附近的松树林之间，并给予了环境和现存的树林最大的尊重。

剧场的朝向打破了历史轴线，与场地的主要构成元素相得益彰。为了维持统一性，并且在葡萄酒庄园内突出新剧院，其结构还要依靠轴线的力量。这一文化场所为其周围环境和过客提供了新的漫步路径，人们可以穿梭其间，同时它还是一处新型聚会场所。规划的广场是公众放松娱乐的聚集地，广场拓展了葡萄酒庄园的空间，且同带有裂口的入口处一样，兼做登记处，同时还是一处等候区和展览区，站在这里，在这片区域的主轴线上闲庭信步的人群被尽入眼帘。

布景效果

所有选用的舞台布景、配景设计和空间设计都通过主要的建筑理念连接在一起。

建筑师的想法是致力于技术追求，如人行交通流线、舞台的能动性、空间利用以及其毗邻的实体建筑。建筑师努力打造体现剧场最佳效果的设计方案，以及研究道具如何运递等。建筑师开发了一套配景设计，它在适应和开展任何文化和艺术活动方面具有巨大的潜力。

节奏与深度

剧场对哈尔昆（Harlequin）织物的隐喻意义非常明显，这也是Domaine de l'O非常珍视的特色。建筑师重新定义了主题，以设计灯光、对比强烈的色彩和纹理。木嵌板在立面上纵横交错，包裹着整个体量。大厅的结构也是由木材建成的，其倾斜性避免了单调，整体活泼，且形成了几种韵律，使这座火红的剧场在必要的时候能为人们所见。

巨大的交错通过明亮的菱形灯的点亮得以加强，这是对哈尔昆品牌的另一个参考。夜间，这些灯光吸引人们的眼球，创造了有趣的氛围。这种交错性还延伸至入口处，使空间最大化，并且在周围提供了一个全新的视点。木质外层将其本身托举起来，并且展示了它的秘密。观察员们被要求入内，去发现一处为剧院服务的空间。夹层允许他们望向其中，同时处于其间的人们也可望见外面的人们。

菱形在室内也有所应用。在入口处，这一木质"外衣"将其本身托举起来，来展现剧院空间的秘密。观察员们在底部滑过，来发现与外表面同等处理的空间。菱形与外部相回应，再现了室内空间。

如果说菱形蕾丝纹饰是剧院的标志，剧院本身也隐藏了许多关于文化、剧场和选址的内涵的其他信息。火红的建筑外壳重新应用了本区域内的赭石，提高了它与场地的和谐性。剧院完美地融入到了葡萄酒庄园中。

环保方面的杰出性能

低耗能建筑：

让·克洛德·卡里埃剧院的耗能比常规建筑减少一半，堪称低耗能典范。剧场无需像传统建筑那样大量消耗能源（供暖、冷却、通风、采光所需的），剧院舞台内使用的所有灯都是LED灯，这在法国尚属首创。动态热模拟的重复使用证实，这些节能是与当地地中海气候的舒适性是相关的，人们在这一气候区是无需配置空调系统的。

可持续建材：

通常，建造一个项目所需的能源大约相当于30—50年资源开采所需的能耗，考虑到力能参数，建筑师决定使用可持续的节能建材。这一新建文化设施完全为木质结构，外墙、屋顶框架、地面、墙体以及外立面展现了一个大约为1000m³的体量。建筑师同样注重从橡胶地板到涂料以及经过认证的玻璃方面的其他选材。最后，剧院内的生命循环分析证明了其仅相当于7年的资源开采年耗。

项目名称：Jean-Claude Carrière Theater
项目地点：Montpellier, France
建筑师：Philippe Bonon, Philippe Cervantes, Gilles Gal
项目经理：Tiffanie Renard
项目团队：Spie-Sud Ouest, A+Architecture, Calder Ingnierie, Structures Bois Couverture, Betem, Arteba, Celsius Environnement
项目工程师：Calder Ingnierie
甲方：Conseil Général de l'Hérault
用地面积：23ha 楼层面积：2,620m²
材料：KLH panels, laminated timber(from Austria) for 650m³ of the wood + 350m² of French wood / coated metal leaves for a waterproof envelope and where it's accessible stained resin panels.
竣工时间：2013.6 造价：EUR 6,687,000
摄影师：©Marie-Caroline Lucat(courtesy of the architect)

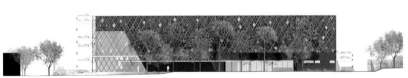

东南立面 south-east elevation

东北立面 north-east elevation

西北立面 north-west elevation

西南立面 south-west elevation

1 夹层	1. mezzanine
2 集气室(望向大厅)	2. plenum(looking onto a hall)
3 集气室(俯视舞台)	3. plenum(overlooking the stage)
4 未折叠的坐椅	4. unfolded seat area
5 淋浴区和艺术家更衣室	5. showers & artists dressing room

三层 third floor

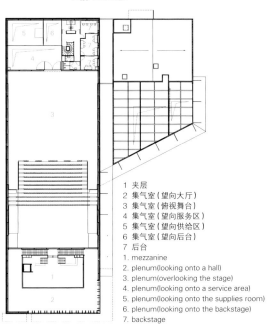

1 夹层	1. mezzanine
2 集气室(望向大厅)	2. plenum(looking onto a hall)
3 集气室(俯视舞台)	3. plenum(overlooking the stage)
4 集气室(望向服务区)	4. plenum(looking onto a service area)
5 集气室(望向供给区)	5. plenum(looking onto the supplies room)
6 集气室(望向后台)	6. plenum(looking onto the backstage)
7 后台	7. backstage

二层 second floor

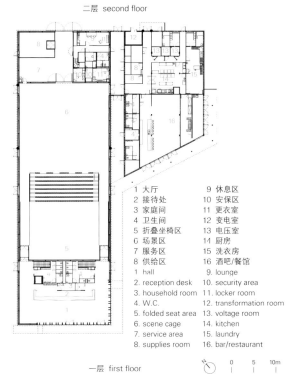

1 大厅		9 休息区	
2 接待处		10 安保区	
3 家庭间		11 更衣室	
4 卫生间		12 变电室	
5 折叠坐椅区		13 电压室	
6 场景区		14 厨房	
7 服务区		15 洗衣房	
8 供给区		16 酒吧/餐馆	
1. hall		9. lounge	
2. reception desk		10. security area	
3. household room		11. locker room	
4. W.C.		12. transformation room	
5. folded seat area		13. voltage room	
6. scene cage		14. kitchen	
7. service area		15. laundry	
8. supplies room		16. bar/restaurant	

一层 first floor

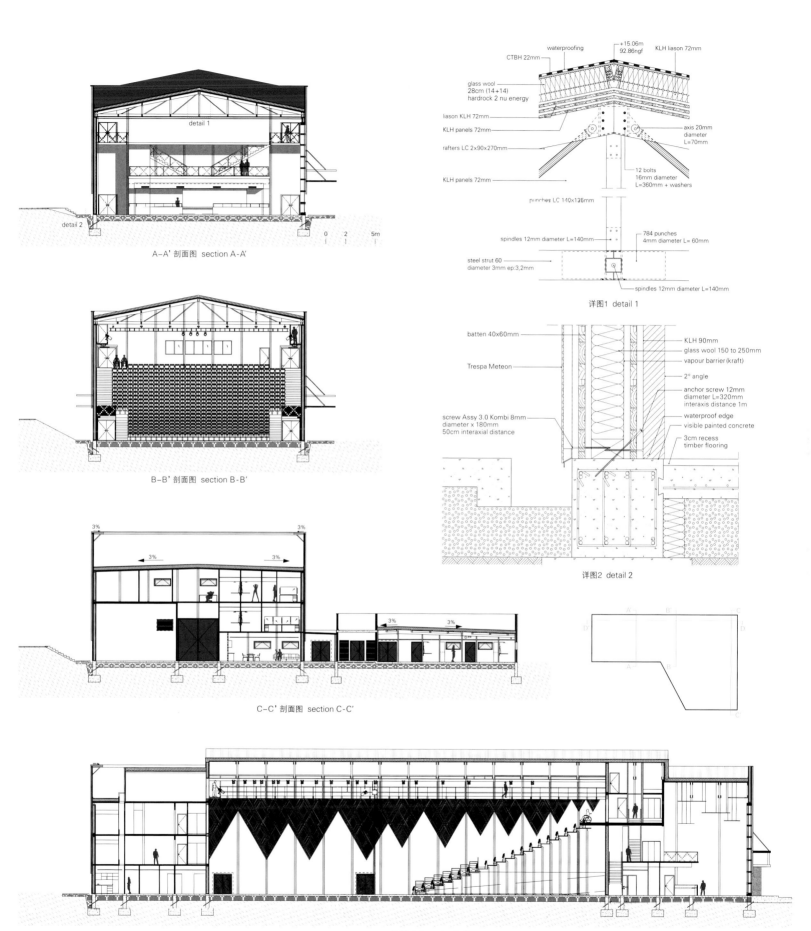

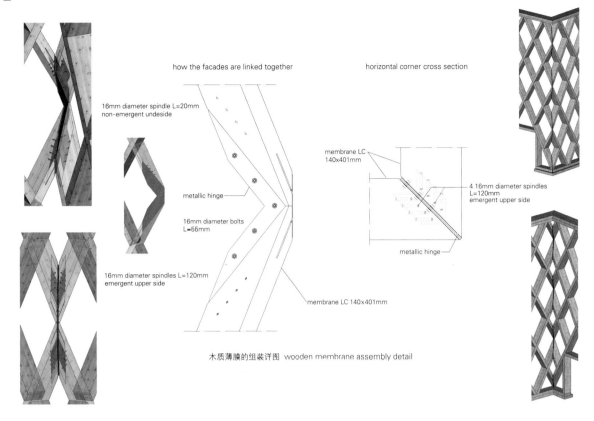

木质薄膜的组装详图 wooden membrane assembly detail

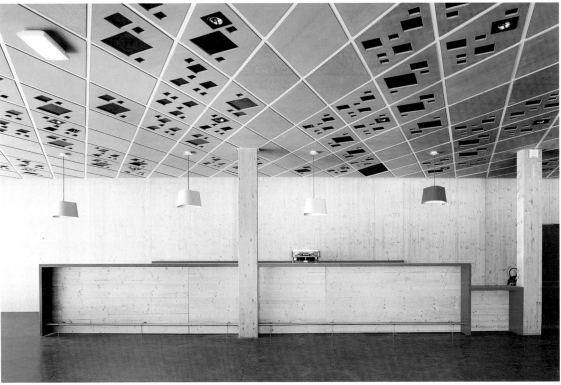

Jean-Claude Carrière Theater

Implementation and Site

The Domaine of O, site for which the architects conceived the project is breathtaking; its forest of pine trees, its history, the space it unrolls inspired a light and delicate concept. The Theater lands between the esplanade set for big tops and the nearby pine trees with the utmost respect towards the environment and existing trees.

The theater orientation both breaks the historic axis and fits in the main composition of the site; its structure relies on the strength of this axis in order to maintain unity and to enhance the new theater in the Domaine of O. This culture box offers its surroundings and passersby new pathways to wander through and new spots for gathering. The proposed square offers a deep breath to the public feed. It extends the domain of O very airy spaces and plays the same register than the entrance with its large crevice. It is also place of waiting and the show, to observe those who wander on the major axis of the field.

Scenography

All choices in terms of stage set, scenography and space designs are linked together with the main architecture concept.

The architects' thinking was devoted to technical requirements such as the pedestrian circulations, the mobility of the stage, the space management and its adjoining entities. The architects wondered the best way to welcome theater ensembles, how their props would be delivered, etc.. The architects developed a scenography that has a huge potential to adapt and open to any cultural and artistic activities.

Rhythms and Depth

The metaphor of Harlquin's fabric is obvious, a character that is cherished by the Domaine of O. The architects redefined this theme to create a light and tangy color and texture. The wood panels cross each other and envelop the volume. Wood structures the hall as well. Its declinations avoid monotony, and they vibrate and create several rhythms and let the red box be seen when necessary.

This vast weave is punctuated with sparks of bright diamond-shape lights, another reference to Harlequin. At night they strike the eye and create an electric and playful ambiance. It is distended on the entrance to magnify this space and offer a new point of view on the surroundings. The wooden dress lifts itself up, and reveals its secrets. Spectators are invited to enter and discover a space dedicated to theater. A mezzanine allows them to see and be seen. The diamond-shapes also echo inside. On the entrance, this dress lifts up itself to let show through the secrets of this place of theater. The spectator skips under these bottoms to discover a space treated with the same care as the outside facing. This diamond is put in echo and repeats in the internal space.

If the lace of diamond-shapes is the signature of the project, it also hides many other references to domains related to culture, theater and to the site of course. The red tint of the building's shell resumes ochres present on the domain; it adds to the architectural coherence of the site. The theater is delicately integrated into the Domaine of O.

Environmental Excellency

A low energy consumption building:

By consuming twice less than a regular building, the Jean-Claude Carrière Theater complies with the Low Consumption Label. The theater does not require much energy as for the conventional (heating, cooling, supply air, lighting), adding to this, all lights on stage are 100% LED, making the project the very first one in France. The iterations of dynamic thermal simulations allowed to verify that these energy savings go hand in hand with the comfort of usage of the Mediterranean climate with no systematic need of air conditioning.

Sustainable materials:

Usually the energy necessary to build a project equals to 30 to 50 years of exploitation. Aware of this energetic parameter, the architects decide to use sustainable and energy efficient materials. The new cultural equipment is entirely made of wood, and the peripheric walls, the roof framing, the floors, the walls, the facade facings represent a volume of around 1000m³. The architects put the same care in the other materials from the rubber floors to the paintings and certified glazings. In fine the theater circle of life analysis shows its exploitation years is of only 7 years. A+ Architecture

艺术画廊与剧院

乌镇大剧院
Artech Architects

宛如一盏并蒂莲,乌镇大剧院盛开于梦境似的古镇水面上……

乌镇是中国南方远近闻名的传统水乡,由于具有浪漫且超现实的氛围环境,这一开发项目的所有者认定乌镇将在全球戏剧节中发挥着重要的一笔,并且将乌镇定为国际重要戏剧节的活动据点,为了实现这个愿景,大元建筑事务所接受委托,来设计这一组剧院空间。而在这一系列的项目内,乌镇大剧院是最重要的篇章。

大剧院位于基地东侧,与游客中心毗邻,剧院为东西走向,所以游客从访客中心开始就其建立一种视觉连接。然而,设计面临的最大挑战是使这座包含1200座和600座的两个剧场的现代剧院在中国江南传统的水乡小镇中和谐共处。剧院采用象征吉祥的"并蒂莲"为形象标志,使两个剧场完美地共享一个舞台区。设计包含两个椭圆形结构,彼此环扣,在形式上一个是透明的,一个是不透明的。环扣的区域高度为30m,为剧院最高点,建筑的最低点仅为10m,且分别向东西方向倾斜。

室内大堂铺有白色大理石,平均高度为16m的室内大堂墙体上装饰了发光的且精致的金箔和银箔,以展现剧院的金碧辉煌,同时与室内墙体中应用的无光泽的粗糙砖墙和木材形成对比。

为了展现当地传统艺术和手工艺的精髓,建筑师精选了当地传统的蓝印花布图案装饰,来作为室内的主题。较大的剧场应用了蓝色和金色,而较小的剧场则应用了红色和金色。所有的色调都为剧场增添了一种华丽感和优雅感。

剧院入口设在建筑的南侧,并设有一个室外广场,游客可以从小岛出发,穿过桥梁,乘坐木船或者步行到达剧院。右侧的小剧院位于"不透明"的体量中,倾斜墙体的踏板形状的部分覆盖在古老的超大型砌砖上,将门厅包裹起来。大型剧院位于左边,被曲折的扇形玻璃围合,玻璃前面带有明显的中国风装饰,在夜晚闪闪发光,并且映射在水面上,为这超然的水上村落的梦幻般氛围增添了另一番风情。乌镇大剧院的设计和施工规划高效有序,建筑师于2010年接受委托,而项目却于2013年顺利竣工,于5月举办第一届乌镇国际戏剧节,被评为"中国最美丽的剧院"。

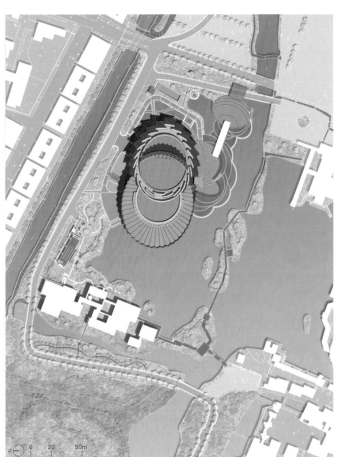

Wuzhen Theater

Like a twin lotus, the theaters rise from the water in this dream-like town...

Wuzhen is one of the most well-known traditional water villages in southern China. In this romantic and surreal environment, the owner of the development decided that Wuzhen would be an important name in the global atlas of theater, and would be where an international theater festival would be held. In order to complete his vision, Artech was asked to design a group of theater spaces. Within this group project, the Wuzhen Grand Theater is the most important piece.

Locating at the east side and close to the tourist center, the theater has an east-west orientation so visitors can have a visual connection to it from the visitor center; however, the greatest challenge of the design was to harmoniously fit this large building containing two theaters with 1,200 and 600 seats back to back, with modern theater functions, in this small, traditional water village setting. Using the culturally auspicious "twin lotus" as its metaphor, which functions perfectly with two theaters sharing one stage area, the design is composed of two oval shapes interlocking one another, one is transparent and the other is opaque in form. The interlocking area of the two oval shapes is 30-meter height, which is the highest point of the building. Slanting to the east and west directions, the lowest point of the building is 10-meter height.

Given their dual purposes of the theater festival and tourism, the functions of the theaters are multiple. Possibilities include formal stage performances, avant-garde creations, fashion shows, conventions and wedding ceremonies.

The interior lobby is paved with white marble. Shimmering and delicate gold and silver foils are used at each side of the average 16-meter high interior lobby walls to show the glamorous of the theaters and also to contrast with the matt and rough brick and wood materials used on the exterior walls.

To reflect the spirit of the local traditional arts and crafts, the flowery patterns of the local indigo print cloth is selected as the theaters' interior theme. For the large theater, the indigo and gold colors are used, while the combination of red and gold is applied

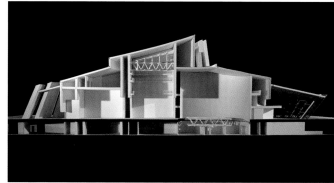

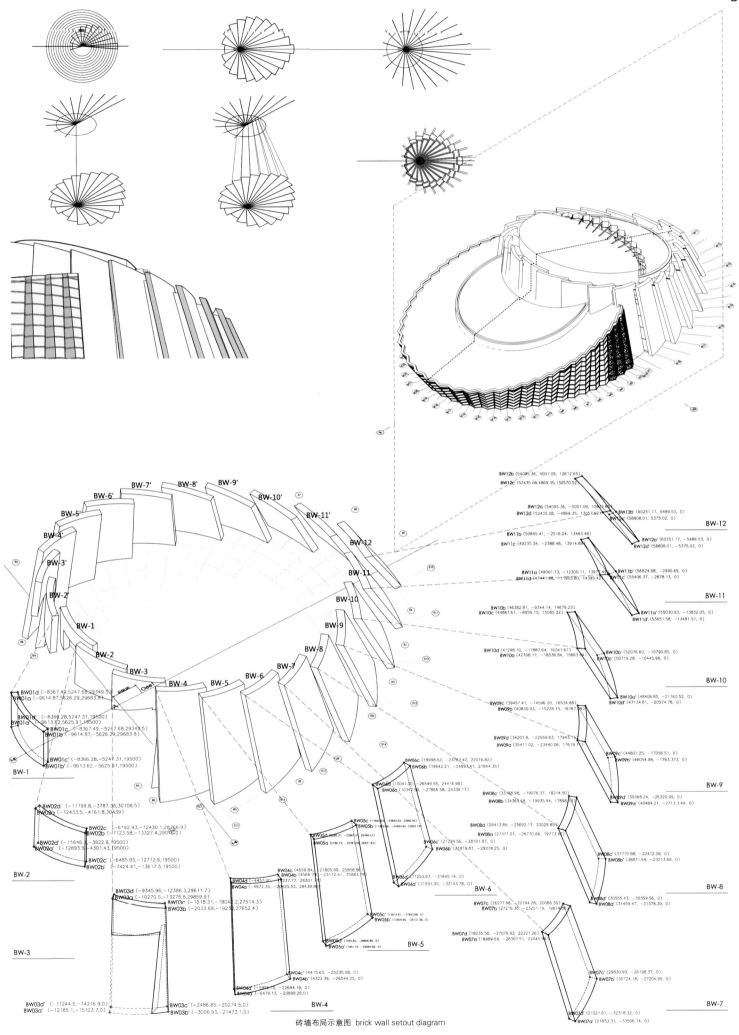

砖墙布局示意图 brick wall setout diagram

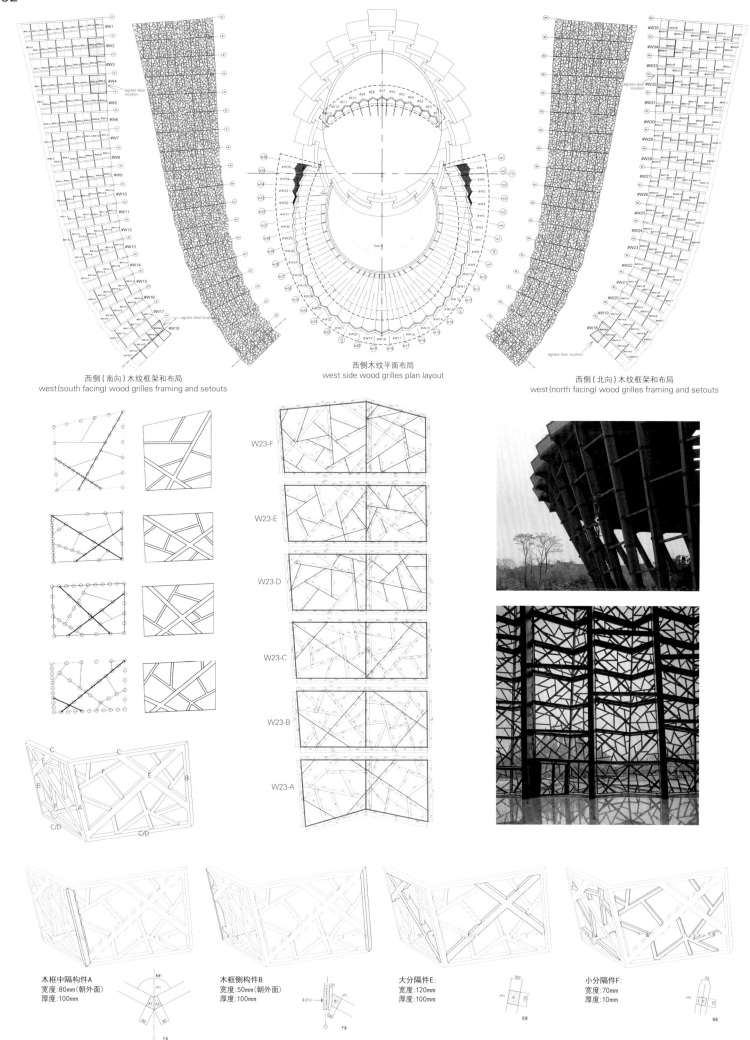

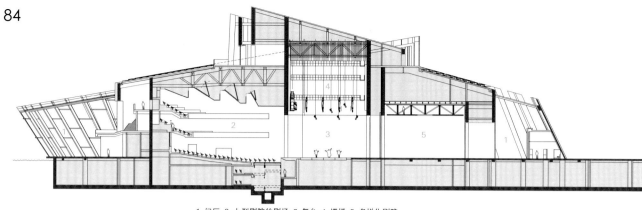

1 门厅 2 大型剧院的剧场 3 舞台 4 塔楼 5 多样化剧院
1. foyer 2. grand theater auditorium 3. stage 4. flytower 5. multiform theater
A-A' 剖面图 section A-A'

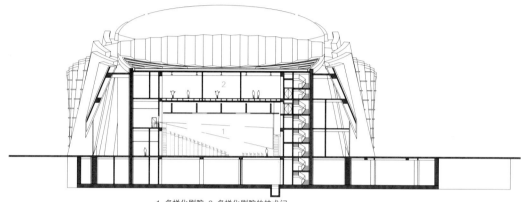

1 多样化剧院 2 多样化剧院的技术间
1. multiform theater 2. multiform theater technical grid
B-B' 剖面图 section B-B'

for the smaller theater. Both palettes provide a gorgeous and elegant sense of the design.

The theater entrance is set in the south side of the building, with an exterior plaza. Visitors may arrive at the theaters by wooden boats or on foot from an island across a bridge. The smaller theater to the right is located within the "solid" volume, where pedal-like segments of thick reclining walls, clad in ancient super-sized brick, wrap around the foyer. The grand theater to the left, enclosed in the zigzag fan-shaped glass front with a Chinese window motif, glows in the evenings and reflects on the water, adding charm to the already misty and surreal atmosphere of this otherworldly water village. Wuzhen Theater project was designed and constructed with an efficient schedule. The design was commissioned in 2010, and the building was completed and opened for the 1st Wuzhen International Theater Festival in May, 2013. The theater has been hailed as "the most beautiful theater in China".

项目名称：Wuzhen Theater
地点：Zhejiang, China
建筑师：Kris Yao
设计团队：Taipei_Kuo-Chien Shen, Winnie Wang, Wen-Li Liu, Jake Sun, Andy Chang, Kevin Lin/Shanghai_Wen-Hong Chu, Fei-Chun Ying, Nai-Wen Cheng, Chu-Yi Hsu, Qi-Shen Wu, Jane jiang
合作设计单位：Shanghai Institute of Architectural Design & Research Co., Ltd
建筑结构：reinforced concrete, steel framing
承包商：Jujiang Construction Group
甲方：Wuzhen Tourism Development Co., Ltd
顾问：theater_Theatre Projects Consultants Ltd / facade_maRco Skin Studio / acoustics_Shen Milsom &Wilke Ltd
用地面积：54,980m² 场地覆盖面积：6,920m² 总楼面面积：21,750m²
建筑规模：2 floors above ground, 1 floor below ground
材料：blue bricks, glass curtain wall, wood grilles
设计时间：2010.5—2010.12 施工时间：2011.1—2013.4
摄影师：
©Jeffrey Cheng(courtesy of the architect) - p.76~77, p.80, p.84, p.86top
©Ying(courtesy of the architect) - p.83, p.85, p.86bottom
©Wuzhen Tourism Development co. - p.78~79

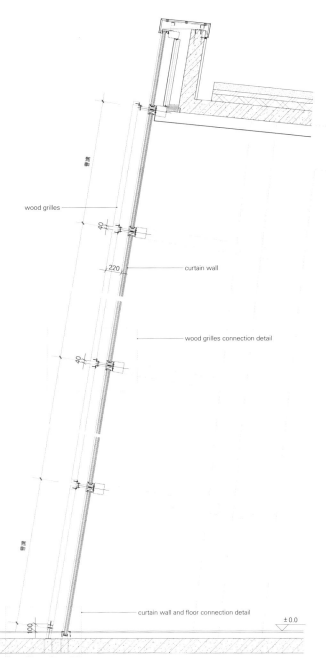

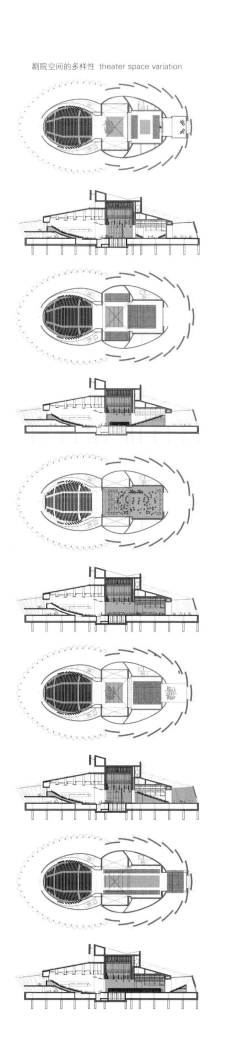

剧院空间的多样性 theater space variation

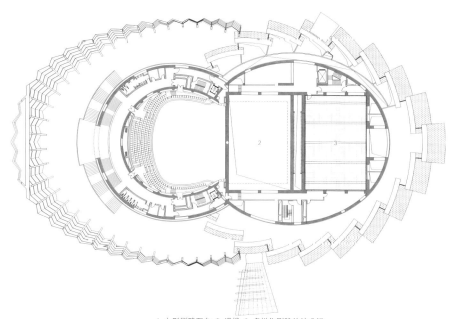

1 大型剧院阳台 2 塔楼 3 多样化剧院的技术间
1. grand theater balcony 2. flytower 3. multiform theater technical grid
三层 third floor

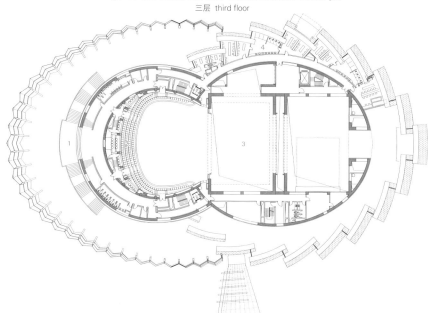

1 门厅 2 大型剧院的阳台 3 塔楼 4 化妆室/后勤室
1. foyer 2. grand theater balcony 3. flytower 4. dressing room/BOH
二层 second floor

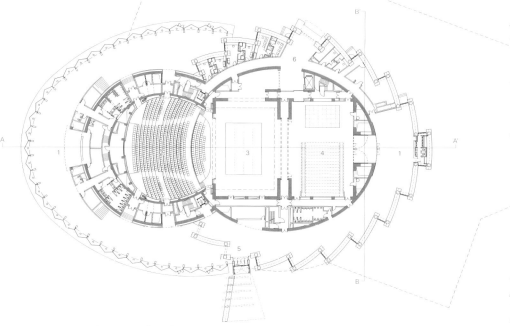

1 门厅 2 大型剧院的剧场 3 舞台 4 多样化剧院 5 入口大厅 6 化妆室/后勤室
1. foyer 2. grand theater auditorium 3. stage 4. multiform theater 5. entrance lobby 6. dressing room/BOH
一层 first floor

艺术馆
Future Architecture Thinking

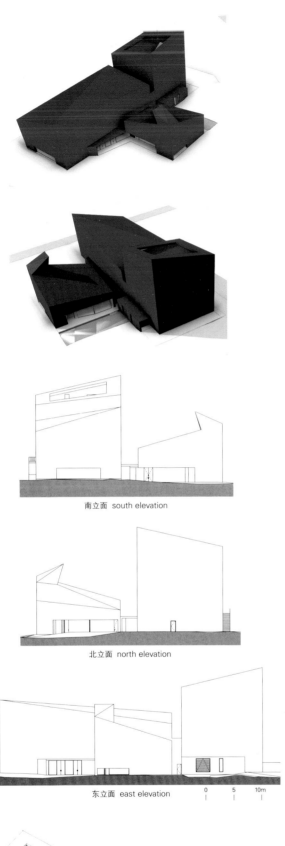

南立面 south elevation

北立面 north elevation

东立面 east elevation

0 5 10m

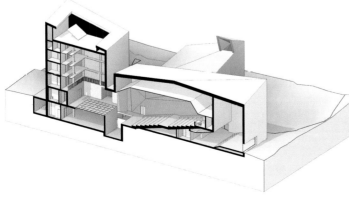

这一项目位于葡萄牙Miranda do Corvo地区Lousā山脉点缀的景观中，彰显了城市与乡村两种特征的交织。

该项目的建筑语言是现代的，且从体量上来说，也是令人印象深刻的。倾斜的屋顶与山地景观的几何外形建立了一种对话，且与周围村庄的屋顶十分类似。立面与屋顶之间的活力通过大红色得以凸显，并且通过周围景观区域内的植被来突出其设计和建筑本身。

艺术馆不仅仅是一座建筑，更可以当做当地的地标，用来标志人们相会、文化与艺术产生的区域，这是一处促进和激发创新性活动、提高人们生活品质的地方。

部署区域进行了优化，便于景观区域的利用，一座用于举办户外活动的圆形露天剧场建造起来，且融入在花园中，花园是村庄共有的一处区域，其中的几处空间以及小径都用作居民休闲之用。

整座建筑包含三个不同用途的体量：第一个体量包含舞台区域；第二个体量为观众席和门厅；而第三个体量设有一间自助餐厅，还有未来建造博物馆的区域，它们在视觉上是一个独立的体量。

多入口的设计试图突出场地作为一处公共空间的特点，同时允许公众无需穿过礼堂便能直接到达特定区域，如博物馆区或自助餐厅。

主入口穿过门厅。门厅可用作展览区，而一条较短的阶梯将门厅一分为二。从这里分出两条路径，通往容纳300人的礼堂，这个礼堂采用机械化控制的乐团席，设有六个技术楼层，来较好地适应如戏剧表演、歌剧、音乐会、会议或讲座活动的需求。

自助餐厅可独立于建筑的其他部分运作，也是进入礼堂的一个入口。自助餐厅顶部为带有一个朝西的天窗的覆顶平台，能够将傍晚的阳光引入室内，平台设有可通往多媒体室的入口。规划中的博物馆区的立面面对花园北侧和室外剧场，花园同时也是建筑的主要入口处所在。

House of the Arts

House of the Arts in Miranda do Corvo expresses the meeting between two identities, rural and urban, in a landscape marked by the Lousā Mountains.

The building features a contemporary and volumetrically expressive language. The sloping roofs establish a dialogue with the geometry of the mountain landscape, in an analogy to the village rooftops. The dynamism achieved through the continuity between facades and roof is accented by a strong red color, emphasizing its design and highlighting the building through the surrounding landscaped area vegetation.

More than a building, the House of the Arts pretends to be an iconic landmark, celebrating the place where people meet, where culture and art happen, a space capable of promoting and stimulating creative activity, increasing the population quality of life.

The deployment area was optimized to favor landscaped spaces, allowing the creation of an amphitheater for outdoor events, integrated in a garden which is a public space for the village, with several spaces and inviting pathways for leisure.

The building consists of three volumes reflecting different sorts of use: the first one containing the stage areas, the second comprising the audience and foyer, and the third with a cafeteria and a future museum area, which constitute a visually independent volume.

The proposed diversity of accesses for the building attempts to emphasize the characterization of this site as a public space, while allowing the public direct access of specific places, such as the museum area and cafeteria, independently, without passing through the auditorium.

The main entrance is through the foyer. This space may function as an exhibition area which can be divided into two by a short flight of stairs. From here it departs two paths to an auditorium for nearly 300 people, with a motorized orchestra pit and six technical levels, properly equipped for holding theater performances, opera, concerts, conferences or lectures.

The cafeteria can operate independently from the rest of the building, or even serve as an entrance point providing access to the auditorium. This space has a covered terrace with a skylight oriented west, channelling sunset light into its interior. The terrace area gives access to a multimedia room. The facade of the museum area is facing the northern part of the garden where one of the main entries is located and the outdoor amphitheater.

项目名称：Casa das Artes in Miranda do Corvo
地点：Miranda do Corvo, Portugal
建筑师：FAT - Future Architecture Thinking
项目团队：Miguel Correia, Cláudia Campos, Sérgio Catita, Patrícia de Carvalho, Miguel Cabral, Margarida Magro, Sara Gonçalves, Telmo Maia, Gabriel Santos, Hilário Abril, José Pico
工程师：José Pico
景观建筑师：Sara Távora
甲方：Municipality of Miranda do Corvo
建造商：TECNORÉM–Engenharia e Construções, S.A.
用地面积：4,790m²
总建筑面积：2,360m²
有效楼层面积：1,435m²
竣工时间：2013
摄影师：©João Morgado(courtesy of the architect)

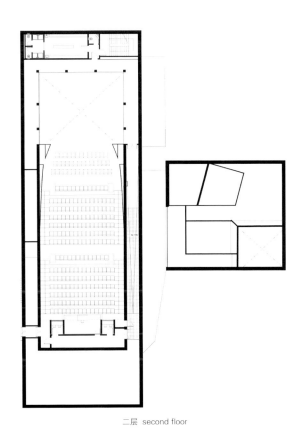

二层 second floor

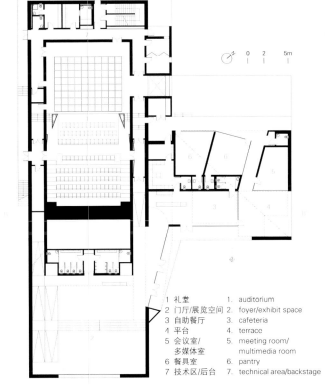

1 礼堂	1. auditorium
2 门厅/展览空间	2. foyer/exhibit space
3 自助餐厅	3. cafeteria
4 平台	4. terrace
5 会议室/多媒体室	5. meeting room/multimedia room
6 餐具室	6. pantry
7 技术区/后台	7. technical area/backstage

一层 first floor

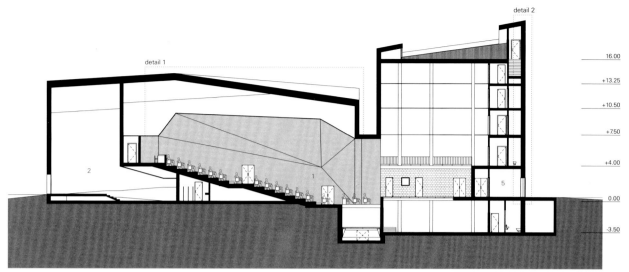

A-A' 剖面图 section A-A'

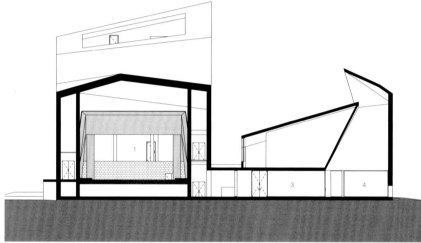

1 礼堂
2 门厅/展览空间
3 自助餐厅
4 平台
5 技术区/后台

1. auditorium
2. foyer/exhibit space
3. cafeteria
4. terrace
5. technical areas/backstage

B-B' 剖面图 section B-B'

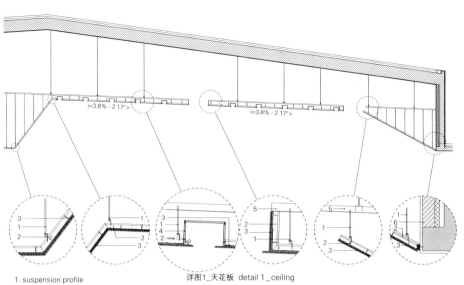

1. suspension profile
2. ceiling structure
3. acoustic ceiling panel
4. lighting box in plasterboard
5. metallic profile
6. inner brick wall

详图1_天花板 detail 1_ceiling

1. roof finishing in zinc painted
2. screw fixation and rubber sealing
3. grid in galvanized steel
4. thermic insulation
5. inner brick wall
6. outer thermic brick wall
7. concrete beam
8. outer birck wall
9. metal clamp

详图2_立面 detail 2_facade

树美术馆
Daipu Architects

艺术画廊与剧院 Art Gallery and Theater

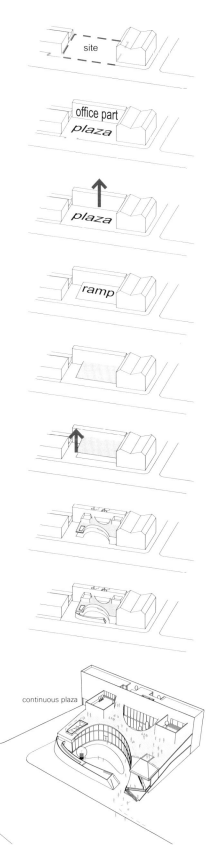

树美术馆坐落在中国北京宋庄,位于该地区一条主公路的路边。原有的村落景观逐渐消失,被大尺度的适合车行的地块划分取代。虽然这里有艺术村的美名,但没有当地艺术家朋友的引荐,人们很难在这一区域停留,对艺术氛围有深入的探访。因此,最早的想法是在基地上创造一处不同于周边环境的,适合人们在这里停留、约会以及交流的公共艺术空间。

建筑师希望人们一开始就被入口处的视野所吸引,视觉能够不自觉地跟随弧形的楼板线进入到美术馆的内部。参观者可以选择从入口处倾斜的楼板进入空间,也可以经由一楼带有水池和树木的庭院进入空间。在这里天空被映射到地面,它与倒影池一起,使人们过滤掉烦恼,也让人忘掉外界的环境。

第一个庭院由一堵暴露在外的混凝土墙体来把外界的马路和灰尘隔离开来。在这里人们可以选择坐在庭院的树下聊天,或是给水池里的鱼儿喂食。同时,人们透过巨大的幕墙还可以欣赏室内的艺术品以及看到人们在室内徘徊。在裸露的混凝土墙体的内部是一个可以用作展览书籍和小型雕塑的走廊。走廊的曲线沿着小径略有不同。

第二个庭院为后面的展厅和二层的会议室提供采光,同时根据公共和私密的需求隔离空间。同时第二个庭院的曲形展墙可以自然地将人们引向另一边的展厅,引向屋顶的公共阶梯式广场。在那里人们可以坐下来晒晒太阳或者俯瞰池塘景色。

场地面积为2695m²,共有六个半庭院。除了展示区的两个,后面的四个半庭院都位于建筑的上部:两个庭院为后部空间提供阳光,同时将阳光引入下方的展厅;另外两个庭院则位于楼顶,面向天空开放。

该项目希望透过真实和纯粹的空间表达,为当地和外来的参观者提供一处与自然光、树、水以及当代艺术交流的空间。这个简单而朴素的想法,将借由参观者的体验向更多的人们传播出去。

项目名称:Tree Art Museum
地点:Songzhuang, Beijing, China
建筑师:Daipu Architects
总监:Dai Pu 项目团队:Feng Jing, Liu Yi
结构工程师:Huang Shuangxi. 机械工程师:Wang Gepeng
电气工程师:Wang Xiang 水利工程师:Lei Ming
幕墙设计顾问:Beijing Doorwin Decoration Co., Ltd
甲方:Chinese Contemporary Art Development Foundation
用地面积:2,695m² 总建筑面积:3,200m²
设计时间:2009.11 施工时间:2010.11~2012.9
摄影师:
©Shu He(courtesy of the architect) - p.94, p.96, p.98, p.99, p.102
©Xia Zhi(courtesy of the architect) - p.100, p.101, p.103, p.104, p.105

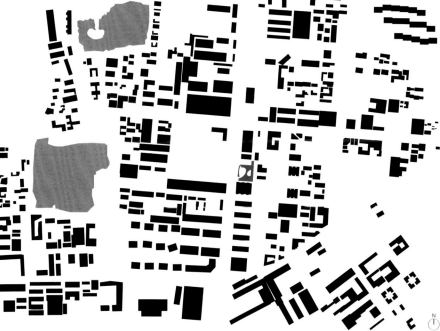

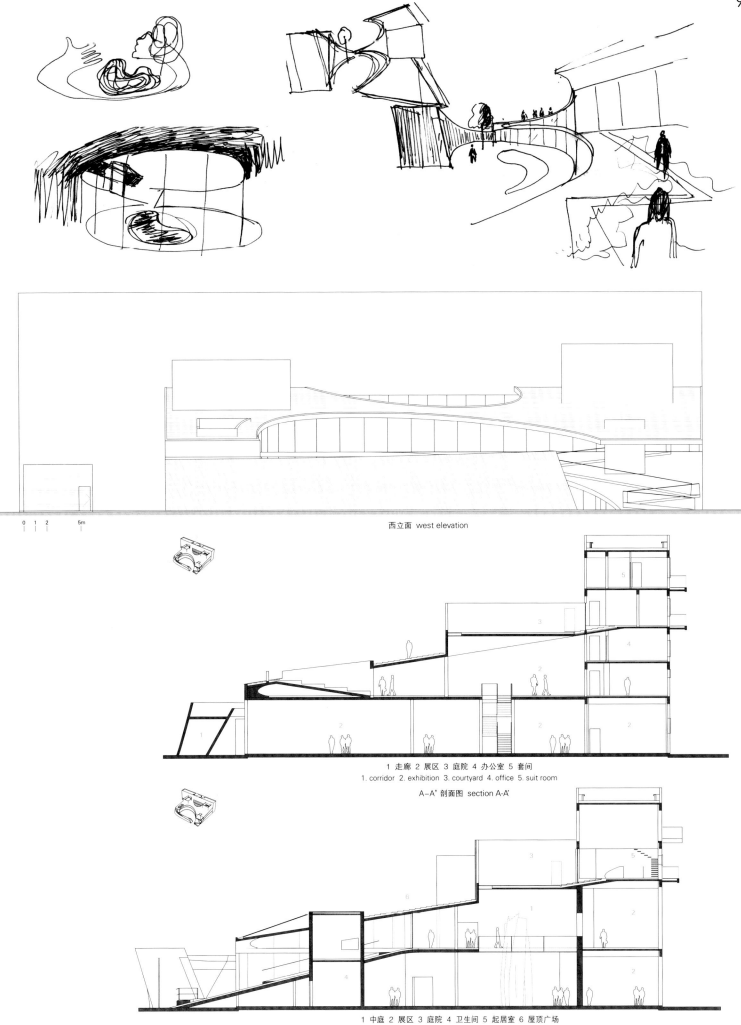

西立面 west elevation

1 走廊 2 展区 3 庭院 4 办公室 5 套间
1. corridor 2. exhibition 3. courtyard 4. office 5. suit room

A-A' 剖面图 section A-A'

1 中庭 2 展区 3 庭院 4 卫生间 5 起居室 6 屋顶广场
1. atrium 2. exhibition 3. courtyard 4. rest room 5. living room 6. roof plaza

B-B' 剖面图 section B-B'

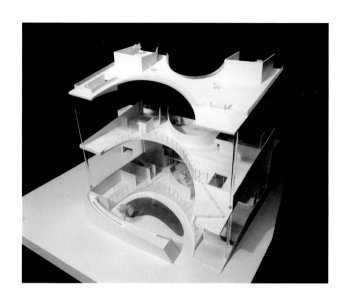

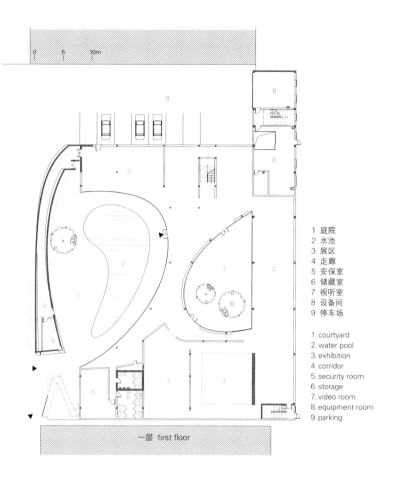

1 庭院	1. courtyard
2 水池	2. water pool
3 展区	3. exhibition
4 走廊	4. corridor
5 安保室	5. security room
6 储藏室	6. storage
7 视听室	7. video room
8 设备间	8. equipment room
9 停车场	9. parking

一层 first floor

1 坡道楼梯/礼堂	1. ramp stairs / auditorium
2 展区	2. exhibition
3 办公室	3. office
4 多功能间	4. multi-function room
5 储藏室	5. storage

1 屋顶广场	1. roof plaza
2 庭院	2. courtyard
3 俯瞰区	3. overlook area
4 起居室	4. living room
5 套间	5. suit room
6 卧室	6. bedroom

三层 third floor

五层 fifth floor

二层 second floor

屋顶 roof   四层 fourth floor

Tree Art Museum

Located in Songzhuang, Beijing China, Tree Art Museum lies beside the main road of the area. The original village has vanished, replaced by big scale blocks which better fit for cars. Even if renowned as artist village, it's difficult to stay or enjoy art exploration without local artist friend's introducing. So the first idea was to create an ambient, public space where people would like to stay, date and communicate.

The architect hoped that people might be attracted into the museum by the view at the entrance. Their eyes would follow the curvy floorslab coming from the ground all the way up to the roof. People could choose getting into the space either through the ramp or the courtyard with a pool and tree on the first floor. Sky is reflected onto the ground, with reflecting pool together, helping people filter their mind and forget the environment out there. The first courtyard was separated with the main road and dust outside by a bare-concrete wall. People would stay and chat under the trees in the courtyard, or, just feed fish by the reflecting pool. Meanwhile, they could enjoy artworks and watch other people lingering inside the building through the curtain wall. In the bare-concrete wall, there is a corridor which could be utilized to exhibit books and small sculptures. The curvature varies slightly along the path.

The second courtyard introduces nature light to the back exhibition hall and meeting room on 2nd floor, while separating the public and privacy needed. The curvy wall implies people to the other side of the building, and introduces them to come to the public stairs-plaza on the roof, where people could enjoy sunshine and have a break, or look down to the pool.

There are six and half courtyards on the 2,695 square meters site. Besides the two bigger ones for exhibition, there are four more courtyards lying on the upper part. Two yards apply sunlight to the back space and introduce skylight to the exhibition hall below. The other two yards are on the top of the floor, which also open to sky. By taking real and pure expression, this project hopes to create a place where local people and visitors would communicate with nature light, trees, water, and contemporary art. This simple and plain idea will spread out through their experience. Daipu Architects

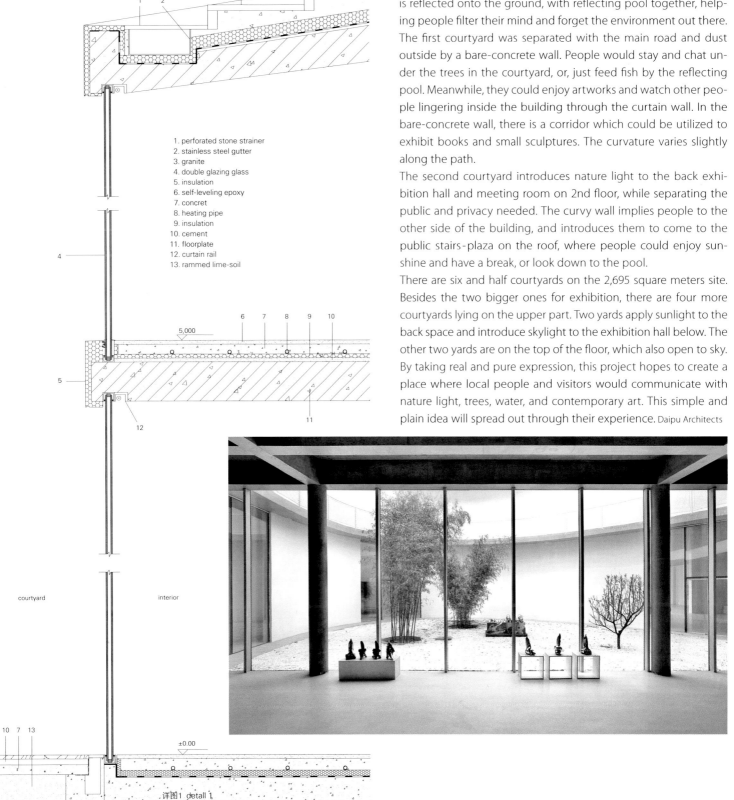

1. perforated stone strainer
2. stainless steel gutter
3. granite
4. double glazing glass
5. insulation
6. self-leveling epoxy
7. concret
8. heating pipe
9. insulation
10. cement
11. floorplate
12. curtain rail
13. rammed lime-soil

详图1 detail 1

丰岛横尾馆

Yuko Nagayama & Associates

坐落于濑户内海的一座古老的港口城市——丰岛,丰岛横尾馆以艺术博物馆的姿态屹立于这个以老年人为主要居住人口的古老社区。

该项目的设计出自艺术领域的国际名人横尾忠则之手,它与由福武基金会支持的直岛博物馆一同作为参与活动的一系列艺术项目。

该艺术博物馆于2013年夏季与2013年濑户内国际艺术祭一同对外开放。该项目起先由三栋一层日式古老房屋构成。对此建筑师更改了设计意图,并对部分建筑进行了翻新和扩建。

博物馆的设计旨在实现由建筑本身就可识别出是横尾忠则的作品的目的。为达到该目的,建筑师首先考虑如何使这两种理念、二维空间和三维空间在几何意义上更加接近。像他的拼贴画作品一样,多种场景和内部体验都被形象地放大并通过每一处屏幕效果、参观者的活动和自然光线的运用形成一幅幅拼贴画。而在该目的的背后,真实的意义却在于强调人类无法避免以及横尾忠则在其作品中一直坚持的"生与死"的主题。"丰岛横尾馆"将会是不断改变自身和在社区中循环的综合体。

其次,该项目能够起到活跃长期生活在丰岛上的老年人生活的作用也同样尤为重要。该作用出自该项目客户的强烈要求。为了与这些老年人分享这一过程和体验,建筑师制造机会让他们知晓该博物馆的目的所在,以及它将如何支持他们的生活。例如,建筑师举行年糕制作

节,并且越过红色玻璃向他们展示施工进程。建筑师和他们一起制作碎瓦,然后将它们铺在花园池塘的底部。正如该馆的主题所表达的那样,该场所确实可以用于举办葬礼。建筑师希望博物馆成为当地人的聚集地,使他们充满活力,并将它尽可能地传播开来。基于这两个前提,建筑师为参观者提供了两条路线,使他们能够在该场地的内部和外部曲折前行。鉴于日式木屋的特色空间,建筑师致力于通过设置黑色透明玻璃和镜面来放大横尾忠则作品的图像和花园的场景。在这些特殊背景当中,红色玻璃的作用实际上是使色彩失去信息的重点,并且该应用使世界的另一端转变成了单一的景色。它同样意味着"生与死"和"普通和特别"之间的界限,而事实上,横尾忠则使用红色作为生命的象征。当人们准备开始博物馆旅程之时,便会发现公园里的红色石头会在视觉上消失不见。而当人们进入内部并穿过主体房屋时,才会遇见本色的花园。从主体房屋望过去,因设计具备不同的功能和不同颜色的玻璃,花园会看起来阴郁一些。而当穿过主楼层下方的溪流时,人们将最终遇见横尾忠则的尺寸为227.3cm×546cm的三部作品——"原始的宇宙"。该作品同样映射在黑色的落地玻璃上,并且能瞬间征服参观者。

项目名称:Teshima Yokoo House
地点:Teshima, Kagawa, Japan
建筑师:Yuko Nagayama, Daisuke Yamagishi
艺术作品和理念:Tadanori Yokoo
施工单位:Naikai Archit Co.,Ltd.
用地面积:444.27m²
建筑面积:184.88m²
有效楼层面积:179.65m²
结构:timber, reinforced concrete
竣工时间:2013
摄影师:©Nobutada Omote(courtesy of the architect)(except as noted)

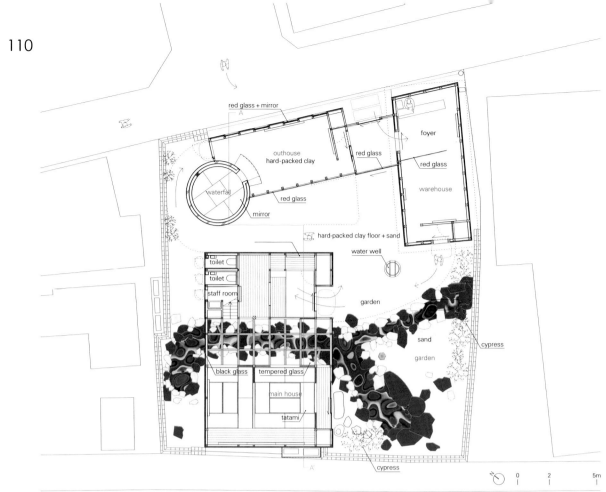

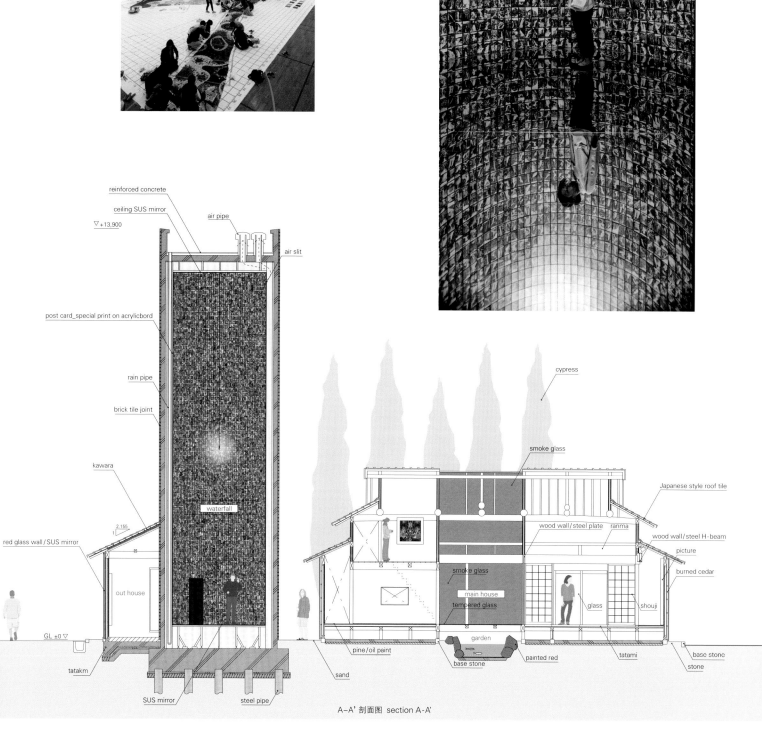

A-A' 剖面图 section A-A'

Teshima Yokoo House

Located in Teshima, an old port town lying in Seto Inland Sea, Teshima Yokoo House sits as an art museum in the old community where most of the people are the elderly.

This is the project by Tadanori Yokoo, an internationally renowned figure in art, as well as being a series of art projects that kicked in with the one in Naoshima, backed by Fukutake Foundation.

This art museum opened in the summer of 2013, aligned with the start of Setouchi Triennnale 2013. It is originally composed of three one-storey Japanese old houses. The architects repurposed, renovated and partly made the extension.

The museum aims at identifying his works with the architecture itself. To achieve this, the architects first contemplated on how they can make close to these two ideas, two dimensions and three dimensions in a geometrical sense. Like his collage works, various scenes and experiences inside are figuratively amplified and collaged by each of screens effects, motion of visitors and natural daylights. Never will you encounter the exact same scene. Behind the purpose, the true meaning underlies that humans can't avoid lifelong theme, "life and death", as he insists through his works. "Teshima Yokoo House" will be the aggregate that is constantly changing itself and circulating in the community.

Secondly, it is also important that this place plays a role in invigorating the elderly people who long lives in Teshima. This is a strong request from the client. To share the process and experience with them, the architects were making opportunity to let them know what this museum is for, and how this place will be side with their lives. For example, the architects held rice cake-

making festival, and showed construction process over the red glass. They actually had done making crushed tile with them to cover the bottom of the pond in the garden. As the theme of the museum says, this place literally is able to hold funeral. They hope this is going to be a hub for the locals, giving birth full of energy to them, and spreading as much.

Based on these two premises, the architects directed new route for visitors to zigzag between the inside and outside of the space. With the characteristic space of old Japanese wooden house, they worked on amplifying the image of his works and scenery of the garden by setting black, transparent glasses and mirrors. Among these special setting, red glass has technically a key to lose information of colors and change the other end of world into monochrome view, this also means the boundary between "life and death" and "ordinary and extraordinary", but in fact, he uses the color of red as an symbol for life. When people are about to start the journey in the museum, they notice the red stones in the garden visually disappear its existence. As they go inside and pass through the main house, they encounter the garden with its original color. Looking from the main house, the garden looked gloomy by distinct functions of different color of glasses. Across the river flowing under the main floor, people'll finally encounter 227.3cm×546cm of his three works, "Primitve universe". This is also reflected in the black floor glass, and overwhelms visitors before they know it. Yuko Nagayama & Associates

建筑师一直致力于在自然资源保护者的要求和新建建筑的功能性需求之间寻求一个折衷方法,而如今这一举动随着(在对历史和文化遗产尊重的前提下产生的)建筑理念而向前迈进了一步。因此,新建结构只能限于建筑表面,因为表面并没有新的考古发现和要保护的结构(即地面)。墙体表面要保持完好无损,或为必要的修复做好准备。

建筑师所要面临的问题是新表面必须将已竣工的空间和即将竣工的空间(好似其起建于不同的时代)连接起来,新表面(如一条毯子)比预期的要展示地大一些。实体设计的空间与黑色混凝土铺路连在一起,这条铺路为中性色调,使其没有与建筑修复部分所产生的复苏之美斗艳,但却足以将等待竣工的混乱部分掩盖起来。这一简单设计的空间理念的特色为主要活动空间(即中殿)的相互交织,在拆除掉增建的楼层之后,接下来的工作便会在现场展示出大量的考古发现。在这一结构部分之上,新楼层在空间内被抬高,形成游客使用的分层式看台,同时还作为重建的巴洛克教堂和之前的哥特式建筑之间的空间分隔。

沿着通往游客座位区而设楼梯的小路为人们提供了一系列不同的空间体验。人们在考古发现中穿梭,站在第一个平台上能够观赏之前的哥特式建筑的楼层平面,而在第二个平台之上,人们能够近距离地观看中世纪建筑新发现的细部,最后,继续向上走,整个巴洛克式中殿的壮观场景逐步展现在人们眼前。新楼层和之前便存在的古老修道院之间的对比在主厅的黑白外观上显得最为突出,并且参考了多米尼克派的黑白服装(白是清白的化身,而黑是谦逊的象征)。

普图伊演艺中心
Enota

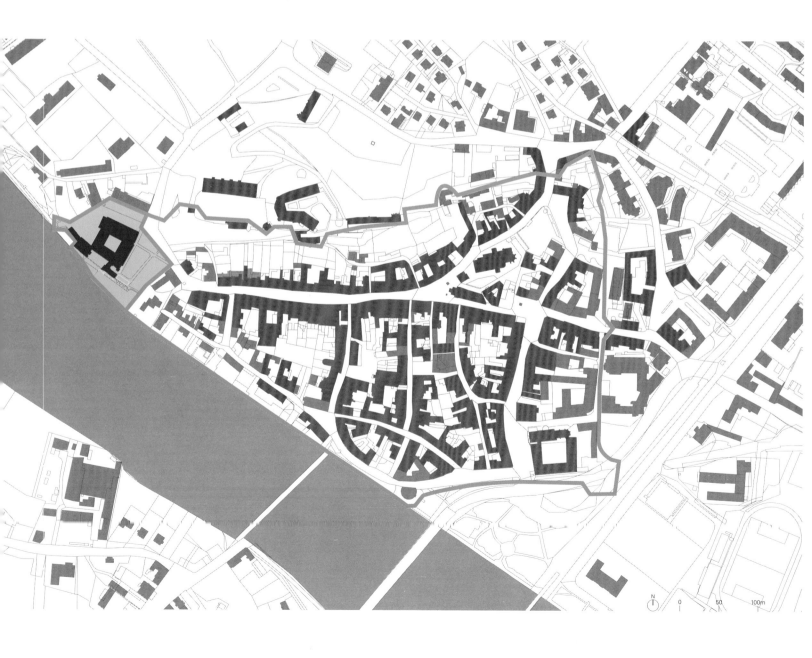

Ptuj Performance Center

The perennial quest for compromise between the demands of conservationists and functional demands of the new content has been taken forward with architectural ideas informed by the attitudes of respect towards the built and cultural heritage. Intervention is therefore limited exclusively to the surfaces where no new archaeological finds or conservation interventions are expected: the floor. The wall surfaces remain intact and ready for the demanding restoration.

The very fact that it has to connect finished spaces and those yet to be finished – originating from different historical periods – the expression of this new "carpet" comes with a little more presence than might be expected. The spaces are tied together into a more solid design whole by black concrete paving, which is sufficiently neutral so as not to compete with the revived beauty of the restored parts of the building, and yet contrasting enough to drown the chaos of the parts of the building still awaiting completion. This spatial concept of simple design features a key twist in the main event space – the nave. After the construction of the added floors had been torn down, subsequent archaeological work re-

vealed rich findings that are presented "in situ". Above this section, the new floor is consequently raised in space, forming tiered stands for the visitors at the same time, and acting as a spatial partition between the reconstructed Baroque church and the remains of the erstwhile Gothic building.

The path along the staircase towards the visitors' seats is thus a sequence of different spatial experiences. Having taken a walk amid the archaeological finds, the first landing offers a view of the floor plan of the erstwhile Gothic building; on the second landing, one can take a close look at the newly discovered details of the Mediaeval architecture; while at the end, towards the top, the view of the entire splendour of the Baroque nave gradually opens. The contrast between the new floor and the historical substance of the erstwhile monastery is deliberately greatest in the black-on-white appearance of the main hall, referencing the famous black and white habit of the Dominican Order where the white is symbolic of innocence and black of modesty.

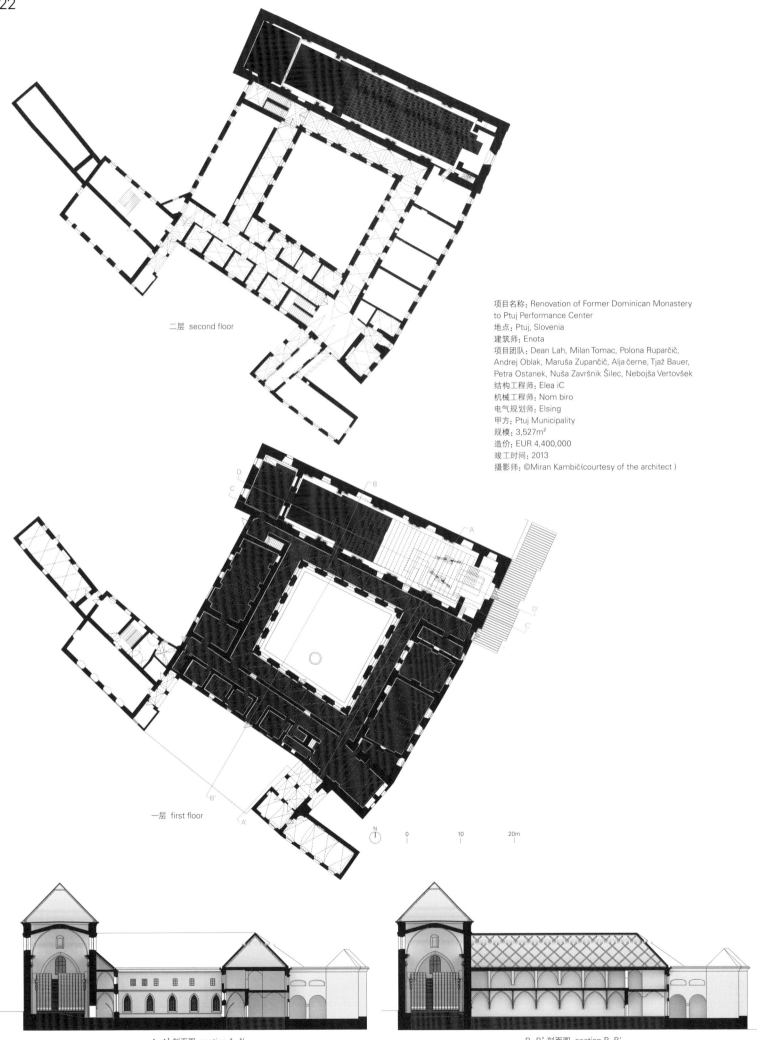

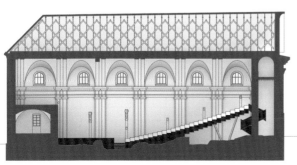

C-C' 剖面图 section C-C'

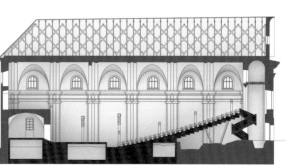

D-D' 剖面图 section D-D'

立面和内部世界
The Facade and

建筑师们通过诸多不同的装置将建筑呈现给外部世界。而在具有物理性质的装置中，立面或"表面"（如果我们参照其词源的话）表示的是位于建筑的内部与外部之间最明显的装置。

几个世纪以来，为了渗透、传播或暗指不同的设计理念，建筑师们已对"表面"的设计进行了处理；这一特殊建筑装置的处理包含一个功能非常强大的工具，因为它在很大程度上能够决定人们对项目的看法，而该看法离建筑内部空间实际性质的远与近成为关键性问题。

在这一点上，本期展示的六个项目，尽管做出向外部世界隐藏建筑内部品质的设计决策背后的原因各有不同，但在设计意图方面似乎存在着共同之处。凭借几种设计策略，建筑师构思出一个双重的解读方法，有意在行人对建筑"表面"的理解与立面背后实际空间之间形成一种差距。

只有当人们在进入建筑之后，才能经历到意想不到的规模变化、建筑语言的改变或内部空间的不同性质。此时，参观者才能最终完全理解该项目的特征。立面因此成为不同视觉和身体感知的物理载体，加剧人们穿过大门进入一个全新和意想不到的世界的信念。

Buildings have many diverse mechanisms through which they are presented to the outside world. Among those of physical nature, the facade or "face" (if we refer to its etymological origin) represents the most obvious device as it mediates between the inside of a building and the outside.

For centuries architects have processed the design of that "face" in order to filter, transmit or imply different ideas; the manipulation of this extraordinary architectural device constitutes a very powerful tool as it has the ability to determine to a big extent people's sight of the project. How close or distant that sight is to the actual nature of the inner spaces of the building becomes the key question.

In this respect, the intention to conceal the qualities of the interior of the building to the outside seems to be common to the six projects herein presented, even though the reasons lying behind that decision are different to each other. Through several design strategies the architects have created a twofold reading, a deliberate gap between passersby's perceptions of the "face" of the building and the actual spaces behind that facade.

Only upon entering the building an unexpected modification of the scale, a change in the architectural language or a different nature of the inner spaces is experienced. Is then when, ultimately, the complete comprehension of the character of the project is achieved. The facade thus becomes the physical embodiment of different visual and physical perceptions, fuelling the belief of entering a gate to a new and unexpected world.

Etymologically, the term "facade" specified the "face" (from the Latin) or principal side of a building presented to the public and containing the primary entrance. This visual and physical architectural device has always represented the main component through which the inside of a building relates to the outside. Through its physical characteristics, the facade regulates the manner and extent of this relationship. Massive and closed facades convey the concepts of defence and impermeability, while those more open and transparent may reveal the inner qualities of the building. The difference between the two types of facades includes a wide range of nuances. Elaborate patterns can create a peek-a-boo effect, subtly confusing the observer about what can really be seen through it by revealing a latticework of lit and shaded areas in the building. This visual relationship between the interior and exterior can also be manipulated through automatic or manual devices, depending on external light and temperature conditions, or uniquely individual choices related to activities and privacy. Projects like Jean Nouvel's Institut du Monde Arabe in Paris or FOA's Housing in Carabanchel (see C3 18) provide eloquent examples of this.

The evocative power of the facade – both as a visual device and a physical and symbolic element – has been employed prolifically throughout the history of architecture. The facade of Renaissance palaces, for instance, was not merely a front to the public space, but, rather, truly a statement of the authority a family held in society. Impressive facades mirrored the wealth and the power of the bourgeoisie living inside. Luxuriously decorated rooms were glimpsed from the street, generating a public image of power and richness. In a wholly unique approach, the architects of the Modern Movement employed the facade to express a new set of societal values, reflecting the ideas of progress, transparency and development. This was possible in part thanks to the progress in construction that characterised the first half of the twentieth century in Europe, which allowed the facade to be freed of its

the Inner World

从词源上讲,"立面"这一术语指代建筑呈现给公众和包含主要入口的"表面"(源于拉丁语)或主墙面。这一视觉和物理建筑装置一直代表着联系建筑内部和外部之间的主要构件。通过它的物理特性,立面掌控着这种关系的方式和程度。宽广和封闭的立面传达防护和抗渗性的设计理念,而开放和透明的立面可能会显露建筑的内部品质。这两种不同形式的立面包含很多细微差别。精心设计的模式能够产生一种瞄孔效果,通过展示建筑中照明和阴影区的网格结构巧妙地使观察者疑惑立面之后究竟会是如何布局的内部空间。这种内部与外部的视觉关系也可以通过自动或手动装置来进行操控,而这取决于外部光线和温度情况,或与活动和隐私性有关的独特选择。Jean Nouvel设计的位于巴黎的阿拉伯学院或FOA设计的位于卡拉班切尔的住宅(参见C3中文版第18期)这两个项目就为此提供了强有力的范例。

由立面(既作为一种视觉装置又作为一种物理和象征性的构件)引起的权力象征,在整个建筑史当中都有丰富的应用。例如,文艺复兴时期的立面,不仅仅作为公共空间的正面,还是一个家族社会地位的真实表现。给人留下深刻印象的立面反映了生活其中的中产阶级的财富和权力。人们从街上瞥见装饰豪华的房间,便产生了一种权力和富有的公共形象。现代建筑运动中的建筑师将一种十分独特的方法运用到立面当中,以表达一套新的社会价值观,反映思想的进步、透明和发展。这一应用得以实现,一部分得益于20世纪上半叶欧洲建造业的发展,是它的发展使得立面脱离了承载功能的枷锁。建筑"表面"第一次从主结构中独立出来,并且能够超出外部支柱或变得完全透明,从而将建筑的内部直观真实地展现出来。

多年来,建筑师们已经学会掌握将立面作为第一眼就可以渗透、传达或暗示建筑信息的强大交流装置的应用。

本期展示的六个项目向我们呈现立面作为向其周遭环境传达和控制项目"表面"的自主构件的应用的根本观点。这些项目似乎都存在一个共同的主题:向外部隐藏内部空间品质的意图。其设计策略的核心便是造成行人感官与立面背后存在空间之间的显著差异。只有当进入建筑后,人们才能够看见意想不到的规模、语言或内部空间性质上的变化。立面成为不同视觉和身体感知的物理载体,加剧人们穿过大门进入一个全新和意想不到的世界的信念。

由建筑师Adamo-Faiden设计的Venturini住宅就对上述关系进行了描述。尽管几十年来它经历了数次改造,但从不同生活空间的布局来看,人们仍可以明显看出清晰的具有布宜诺斯艾利斯市中心街区特点的空间组织原则。该住宅面向街道一侧,即装饰优雅的前部由尊贵的所有者居住,而附属建筑则设计用于出租用途。

该住宅原始部分之间存在的这种潜在差异因新创建空间所应用的建筑语言的转换而变得更加明显。从建筑外部看,只有装饰檐口上方的上部扩建物的暗示能够引发外来者感觉内部产生了新空间。而一旦进入到新空间,截然相反的抽象化体量构成、开口分布和具体的整体形象呈现出了一个完全不同的体验。冷材料、内置构件和完全不带任何装饰的设计赋予院落后部的新空间一种完全不同的氛围,从而与住宅前部堂皇而美观的艺术装饰产生了明显的对比。

这种差异在新加坡的幸运书屋项目中显得更为重要。当地法规呼吁保护和修复位于场地前部的原有建筑,同时给予建筑师充分的设

load-bearing function. For the first time, the "face" of the building became independent from the main structure and was able to transcend the exterior pillars or become completely transparent, revealing the inside of a building in a direct and honest way.
Over the years, architects have learned to master the use of the facade as an extremely powerful communicative device that is able to filter, convey or imply messages upon the first sight of a building.
The six projects presented herein offer an ultimate view of the use of the facade as an autonomous element that mediates and controls the "face" of the project towards its surroundings. A common theme appears to exist among these projects: the intention to conceal the qualities of the interior space to the outside. A significant discrepancy between passersby's perceptions and the existing spaces behind the facade lies at the centre of this design strategy. An unexpected change in scale, language or nature of the inner spaces is revealed only upon entering the building. The facade becomes the physical embodiment of different visual and physical perceptions, fuelling the belief of entering a gate to a new and unexpected world.
Venturini House, signed by Adamo-Faiden, depicts the above described relationship. Although it has undergone several transformations over the decades, the arrangement of the different living spaces still leaves visible a clear organizational principle characteristic of the inner-city blocks of Buenos Aires. The elegantly decorated front part of the house facing the street was occupied by the stately owner, while the rear annexes were designed for rental.
This underlying difference between the original parts of the house has been made more explicit with a swift in the architectural language employed in the newly created spaces. From the outside, only the inkling of an upper extension above the ornamented cornice leads outsiders to intuit the new sphere created inside. Once you enter the new spaces, a radical opposite abstraction in the composition of the volumes, openings distribution and over-

幸运书屋/Chang Architects
Binh Thanh住宅/Vo Trong Nghia + Sanuki + Nishizawa
Venturini住宅/Adamo-Faiden
Alcobaça住宅/Aires Mateus
103住宅/Marlene Uldschmidt Architects
Flynn马厩改建住宅/Lorcan O'Herlihy Architects

通往未知世界的大门/Marta González Antón

Lucky Shophouse/Chang Architects
Binh Thanh House/Vo Trong Nghia + Sanuki + Nishizawa
Venturini House/Adamo-Faiden
Alcobaça House/Aires Mateus
103 House/Marlene Uldschmidt Architects
Flynn Mews House/Lorcan O'Herlihy Architects

Gate to an Unexpected World/Marta González Antón

计自由，以在空旷细长的场地后部建造新扩建物。因此，张德昌建筑师事务所保留了面向街道的现有建筑物，同时将建筑内原有的承重砖墙和木质地板显露出来。该建筑外壳成为新空间的载体，而两者之间的分离明确了它们的不同起源，从而体现建筑的历史特征。建筑师考虑通过一个更加引人注目的设计方法，以宽5m、深30m的空旷狭长区域的形式处理场地后部新建建筑的问题。五个相互连接的体积出人意料地组合在一起，并占据了该场地的绝大部分空间。与现有建筑相比，在一层开发设计的新馆则发挥了一个自主的作用；其场地内的体量构成、几何结构和嵌入物都遵循着自己的原则。

结合了花的图案和其他当地原有的临街立面细节的经典建筑构件丝毫没有暗示内部产生的空间、几何结构、材料和氛围。不论是现有体积后部的垂直方向或是沿新体积后部的水平方向，令人惊叹的空间性和显著的空间延长率都不是传统立面的轻柔节奏和规模比例所能比拟的。如今，建筑正面代表的是一个不透明的隔屏，在它的后面可以明显看出新建筑引人注目的效果。

位于越南胡志明市的平盛区某住宅是由Vo Trong Nghia建筑师事务所和Sanuki+Nishizawa建筑师事务所联合设计的。该项目展示了由预制混凝土花墙包覆的三个悬浮体量设计。从外部看去，夹层产生的空间看起来像是露天阳台，能够实现建筑完全的光线渗透性。人们可能会误以为所有的居住空间都封闭于混凝土阴影体积当中，但当你进入该住宅时，你就会改变这样的看法。看起来好像多叶植物占据了整个室外平台，实际平台上还设置了带有玻璃门的宽敞起居室。而该处使用的粗制材料，尤其是成型天花板应用的模铸混凝土，混淆了人们从外部望过来的第一印象。所谓的渗透性更像是空间、应用、材料和光线之间更加丰富的相互作用。此外，探求空间组织上的误解决定了对该项目布局的理解。

位于葡萄牙阿尔加韦的103住宅阐述了一个不同的观点。由于处在两栋传统渔家住宅夹缝的中间，房屋的正面宽度有限，因此引导Marlene Uldschmidt建筑师事务所在内部开发这一新项目。建筑师无疑已充分利用该项目先前的不利设置，凭借占据的空间来充分利用倾斜和不规则地势，而这种空间占据能够允许不同封闭房间的延伸超出室内的限制。几处沿向上地势设计的露台延伸并加强住宅不同部分之

all materialization suggests a complete divergence in the experience. Cold materials, in-built elements and the complete absence of any kind of decoration confer a radically different atmosphere in the new spaces in the rear of the compound. The contrast with the traditionally conceived, stately spaces of the front, whose aesthetics are rooted in art-deco, is noticeable.

This difference is more significant in the Lucky Shophouse project in Singapore. Local regulations called for the conservation and restoration of the original building situated in the front part of the plot, while leaving substantial freedom for the architects to erect new volumes in the empty and elongated back plot. Hence, Chang Architects retained the existing structure facing the street while revealing the original brick load-bearing walls and timber floor slabs inside the original structure. This shell became the container for the new spaces, and a detachment between both clearly delineates the different origins of the interventions that characterise the history of the building. A more dramatic approach has been considered while solving the newly created volume at the rear of the plot, in the form of a vacant elongated area that is five meters wide and 30 meters deep. An unpredictable composition of five chained cubic volumes colonizes the vast majority of this space. The new pavilion, developed entirely on the ground floor, assumes an autonomous role when compared with existing structures; its volumetric composition, geometry and insertion within the plot follow their own principles.

Nothing of the classical architectural elements combined with floral motifs and other vernacular details of the original facade to the street hints at the spaces created inside, nor the geometry, materiality and atmosphere. The imposing spatiality and the pronounced space elongation – whether vertical at the rear of the existing volume or horizontal along the new volumes at the back – are far from the gentle rhythm and human scale proportions of the traditional facade. This front now represents the opaque screen behind which the dramatic effects of the new architecture are noticeable.

Binh Thanh House, designed by Vo Trong Nghia Architects and Sanuki+Nishizawa Architects in Ho Chi Minh City, Vietnam, displays three floating volumes fully wrapped by pre-cast concrete pattern blocks. From the outside, the spaces created in between seem to be open-air terraces which allow full permeability through the entire depth of the volume. One may consider that all of the inhabited spaces are enclosed in the concrete shaded volumes, but once you enter the house, that perception changes. What seemed to be full depth outdoor platforms colonized by leafy vegetation are instead wide living rooms with glazed fronts. The rough materials here used, especially the cast concrete for the shaped ceilings, obfuscated the first impression from the outside. The alleged permeability is instead a much richer interplay of spaces, uses, materials and light. Moreover, a sought misreading of the spatial organization determines the comprehension of the project.

103 House in the Algarve, Portugal depicts a different perspective. Squeezed between two traditional neighbourhood fishermen's houses, the limited front width led Marlene Uldschmidt Architects to develop the new project inwardly. The architects have certainly

间的视觉连接,以及最终由高层至地平线之间的视觉连接。该项目形成的丰富的透视法和多样的空间性无疑与临街空地明显的紧窄感大相径庭,并且增大的生活空间已远超出向公众暗示的狭小体量。

有关该讨论的另一个有趣的观点可以从都柏林(爱尔兰)Flynn马厩改建住宅的设计中得出。通过设置一个全新的体量作为初始立面,将立面作为建筑内部与外部实体场景之间的物理限制的概念推向了极端。因此,Lorcan O'Herlihy建筑师事务所设计的新扩建物隐藏了该建筑后身现存的19世纪马车房。人们从住宅新正面可以进到入口庭院,其具有抽象造型的体量被严格地分成了四部分,展示与原始建筑截然不同的当代语言和空间性。人们一旦穿过第一栋建筑的立面,内部空间的真实特征便会呈现在眼前。错综复杂的高度交织在先前精心设计的建筑正面周围,而增设的玻璃桥连接两个彼此分开的结构。新建筑的外部景色对理解该项目内部的实际开发并没有提供充足的信息,而能够记住原始建筑和其立面的那些人可能会对完成的设计方案有一个更好的理解。

体现对临街空地交流能力的微妙应变力的设计在艾利斯•马特斯设计的位于葡萄牙的Alcobaça住宅项目中得以实现。先前的两层立面,改造前呈现出典型葡萄牙风格形象的住宅(设计有斜屋顶轮廓),经历了一次彻底的转变。应用统一的白色抹灰层将原建筑构件遮盖住,使它们呈现为一幅白色的画布,建筑师就在这块白色画布上完成了他们的设计。艾利斯•马特斯选择将覆盖层下面的原始开口分布展露出来。因此,现有立面和新的细长深开口的几何结构就相互重叠起来。

与前面几个项目中新临街空地隐藏完全不同的新内部空间不同,在该项目中,建筑师提供了一些有关内部空间的暗示,有意为立面营造一种双重解读的效果。新住宅的特性来源于两种不同的性质,分别是出自于现存结构和新创建空间。一旦你进入该项目住宅,一个出人意料的垂直空间、意想不到的深度和外部空间的创新布局就会呈现在眼前,这在外部是无论如何也无法预料到该体验的特性的。不过,建筑师也在新立面的设计上为内部空间提供了一些暗示。

made the most of this formerly adverse disposition; the sloping and irregular topography has been fully exploited in the virtue of a spatial colonization that allows for the extension of the different enclosed rooms beyond their indoor limits. Several terraces climbing up the terrain stretch and enhance the visual connection of the different parts of the house and, ultimately, from the upper levels to the horizon. The rich scenography and varied spatiality created in the project certainly diverge from the apparent constriction of the street front, and the dilated living spaces expand far beyond the small volume suggested to the public space.
Another interesting point of view in this discussion is visible in the Flynn Mews House in Dublin (Ireland). The notion of facade as physical limit between the inside of a building and the physical context outside is pushed to its extreme by interposing a complete new volume as a sort of preliminary facade. As a result, the new extension by Lorcan O'Herlihy Architects conceals the existing nineteenth-century coach house behind. The new front to the entry courtyard, an abstract-shaped volume strictly divided into quadrants, displays a contemporary language and spatiality that is radically opposite to that of the original building. Once the first building-deep facade has been overstepped, the actual character of the spaces inside is unveiled. Various intricate levels intertwine around the former elaborate front, while an additional glass bridge connects the two separate structures. The new outside view of the building does not provide enough information for people to understand how the project actually develops inwards. Those who remember the original building and its facade may better fully comprehend the design approach accomplished. A more subtle strain to the communicative ability of the street front is accomplished with the Alcobaça House in Portugal, signed by Aires Mateus. The original two-storey facade, which prior to the reform offered a stereotypical image of the Portuguese house (with its pitched roof silhouette), has experienced a thorough transformation. A uniform white plaster layer has been employed to cover all of the original elements, which were transformed into a white canvas upon which the architects completed their design. Aires Mateus opted to make visible the original openings distribution under the covering stratum. As a result, the geometry of the existing facade and the new slender, deep openings overlap.
Unlike previous projects in which a new street front hides completely the radically different new interior spaces, here, the architects have provided some clues, intentionally creating the effect of a double-reading on the facade. The identity of the new house is derived from two different natures, that of the existing structure and that of the recently created space. Once you enter the project, an unforeseen vertical spatiality, an unexpected depth and a novel arrangement of the outdoor spaces are unveiled. It is not possible to anticipate the character of this experience from the outside. Nevertheless, the architects have provided a glimpse of what is to come in their design of the new facade.

Marta González Antón

幸运书屋
Chang Architects

幸运书屋位于新加坡加冬区，即加冬区和如切区的附属居民区内保存下来的书店区域。这一区域的保护原则要求项目的前半部得以保留，且进行重修，而后半部分可以重建，并扩建为四层。鉴于幸运书屋的扩建，该区的保护原则要求任何的建设都可以转为住宅性项目。

幸运书屋建于20世纪20年代，此建筑原为一家名为"幸运书屋"的书店，一楼是零售区，楼上分隔为若干间储藏室。

书屋后有一条窄长的水泥空地，周围是3~4层的酒店式公寓和半独立房，其侧面面向一条后巷，巷子两侧排列着居民楼。

新屋主是工作在国外的夫妻，打算回新加坡居住，夫妻二人都是在如切区长大，在加冬路买下书屋，以做归国居住之用，这是一处他们重温儿时记忆的场地。

这对夫妻买下了书屋及屋后空地，打算与他们的建筑师朋友一起，将书屋修整为住宅，屋后的水泥空地改造为花园，并且在花园里扩建单层的房屋。项目书要求空间具有灵活的使用性，作为父母和亲戚来访之用。

对原书屋的前部进行保留宛如在考古工地工作，项目书的一部分都在要求保护原书屋的痕迹，即重新发现、展示并且保护原始结构、装饰以及细部。

巧合的是，这对夫妻还记得儿时成长的岁月里光顾的"幸运书屋"，看漫画书、买音乐录音带以及储备廉价口香糖的成长时光。

建筑师小心翼翼地剥除书屋正立面的多层涂漆，恢复老店的原始色调，以透明的密封层对其加以保护，以防止表皮剥落。书屋前柱上斑驳退色的标志"幸运书屋"得以保留，提醒人们这里曾经是一座书屋。

书屋内部的非承重隔墙被拆除，使人们能够更好地欣赏空间、旧砖墙、木橡及楼板梁，它们经过了精心的修复、清理以及保护。墙上保留的一排孔洞暴露在外，留给人们猜想原来空间布局的线索，且表面楼板梁曾经支撑着一个夹层（作为额外的存储空间）。

建筑师在此区域增设了柱子，支撑覆盖餐厅区域的扩建的屋顶，这些柱子与砖墙表面是分开的。新老砖墙之间的分界面十分明显，以勾勒出原有共用墙的轮廓，老边界墙的零碎部分也保留下来，作为对原始场地布局的借鉴。

中央的花园通往书屋的后部，那里新建了一排单层房屋，建筑师决定将房屋设在地势较低的场地上，且密集地建造，同时将其地面抬高，这一决定带来了绝佳的新鲜空气，同时使房屋的绿化空间和周围环境更舒适。

房屋的后半部分的规划要求其后退距离与每侧边界相隔1m。这一要求保证了书屋前部的布局，且人能一直观看到房屋后部的绿色植物。

两侧的边界墙不是平行设置的，而是呈锥形，一系列的房间间断性地排列，这种交错性形成了垂直空间，以用于对流通风、采光以及通往室外和绿化区的入口。

最终的改造为这对夫妇找回了美好的童年时光——那个时候，人们居住在一个大社区内，每一家的社交空间都是相互联系的，空间简单又实用，日常生活的琐碎被建筑丰富起来。

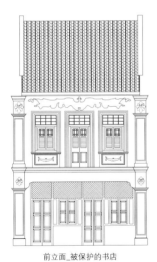

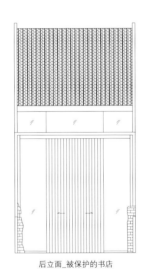

前立面_被保护的书店
front elevation_conserved shophouse

后立面_被保护的书店
rear elevation_conserved shophouse

前立面_扩建的房屋
front elevation_extended house

后立面_扩建的房屋
rear elevation_extended house

Lucky Shophouse

This shophouse is located along Joo Chiat Place, in a conservation shophouse district in the secondary settlement area of Joo Chiat and Katong in Singapore. The conservation guidelines for this area require the front portion to be conserved and restored, while the rear portion can be redeveloped to a maximum of 4 stories. For this stretch of shophouses, the guidelines require any new redevelopment to be converted for residential purposes.

Built in the 1920s, this shophouse used to be a book shop called the Lucky Book Store. The ground floor was the retail area, while the upper level was partitioned for storage.

Behind this shophouse was a long, narrow concreted vacated land. Surrounded by 3 to 4-story service apartments and semi-detached houses, a portion of its side faces a back lane, franked by houses on both sides.

The client is a couple, both have worked overseas and had plans to move back to Singapore. Both grew up in this Katong area and buying a shophouse in Joo Chiat was in many ways homecoming for them. This is a site where they could see their childhood days relived.

And so they purchased this shophouse, along with the rear vacant land. Together with their architect friend, the plan was to convert the shophouse into a dwelling place, and to transform the concrete land at the rear into a garden where a single-story house extension sits. The brief called for flexible usage of spaces, for visiting parents and relatives.

Conserving the front shophouse resembled working on an ar-

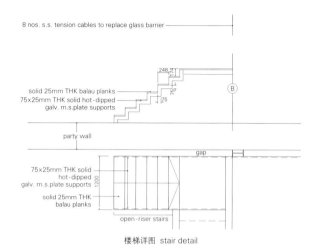

楼梯详图 stair detail

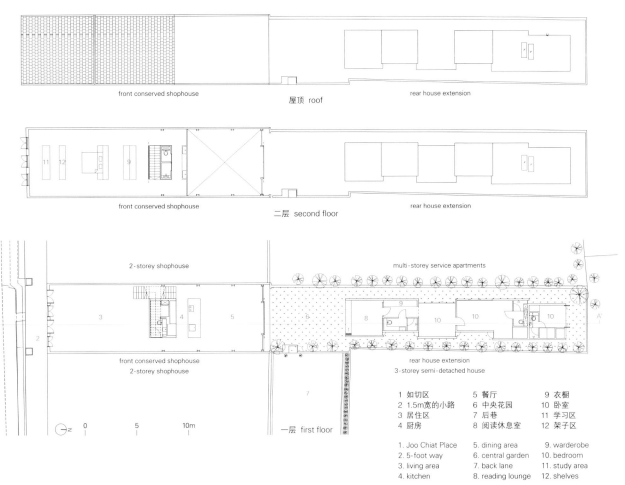

1 如切区	5 餐厅	9 衣橱
2 1.5m宽的小路	6 中央花园	10 卧室
3 居住区	7 后巷	11 学习区
4 厨房	8 阅读休息室	12 架子区
1. Joo Chiat Place	5. dining area	9. warderobe
2. 5-foot way	6. central garden	10. bedroom
3. living area	7. back lane	11. study area
4. kitchen	8. reading lounge	12. shelves

项目名称：Lucky Shophouse
地点：Joo Chiat Place, Singapore
建筑师：Chang Architects
结构工程师：City-Tech Associates
景观专家：Greenscape Pte Ltd
甲方：Yang Yeo & Ching Ian
用地面积：331.20m²
总建筑面积：227.53m²
有效楼层面积：301.36m²
竣工时间：2012
摄影师：
Courtesy of the architect-p.133 middle, bottom
©Albert Lim K.S(courtesy of the architect)-p.133, p.134, p.136, p.137
©Invy & Eric Ng(courtesy of the architect)-p.128, p.129, p.130, p.131, p.132, p.135

1 学习区	6 厨房	1. study area	6. kitchen
2 卧室	7 餐厅	2. bedroom	7. dining
3 衣柜	8 中央花园	3. wardrobe	8. central garden
4 浴室	9 阅读休息室	4. bathroom	9. reading lounge
5 起居室	10 浴室/衣柜	5. living room	10. bathroom/wardrobe

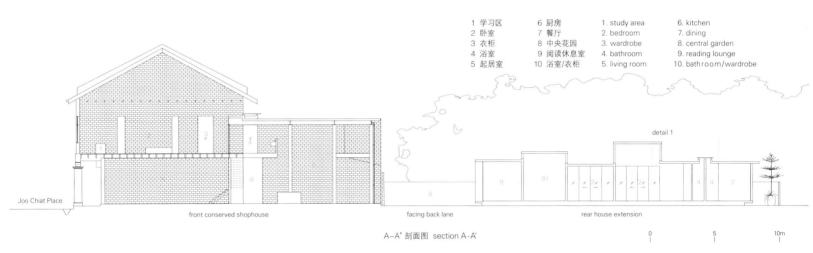

A-A' 剖面图 section A-A'

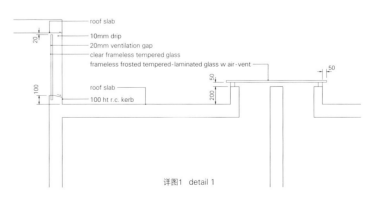

详图1 detail 1

chaeological site. Part of the brief was to retain traces of the old shop, to rediscover, reveal, and protect the original structures, finishes, and detailing.

Coincidentally, the couple remembered patronizing the Lucky Book Store during their growing-up years, browsing through comics, buying music cassettes, and stocking up on cheap chewing gum.

For the front facade, the multi-layered paint-coatings were carefully removed to reveal its original tone and color, and protected with transparent sealers to prevent the surfaces from flaking. The fading signage "Lucky Book Store", spotted on a front pillar, was retained as a reminder to what this place was.

Internally, non-structural partitions were removed so that the spaces, the old brick walls, timber rafters and floor joists, can be better appreciated. These were carefully restored, cleaned and protected. A row of cavities on the walls is left exposed to provide clues of how the spaces were once configured, indicative of floor joists supporting a mezzanine for additional storage space.

New columns are added to support the extended roof over the dining area, these are detached from the surface of the brick walls. Interface between the old and new brick walls is made distinct to reveal the old party-wall profile. Fragments of the old boundary

walls were also retained as a reference to the original site configuration.

The central garden space extends to the rear where the new single-story house sits. The decision to go low dense and to elevate its floor from the ground pleasantly increases the breathing and green spaces for this house and its surrounding.

The planning guidelines for the rear house require a 1m setback from each side boundary. This had also prompted the layout of the front shophouse in such a way that the vistas always open up to the greeneries of the rear house.

As the two side boundary walls are not parallel but tapered, a series of rooms are organized intermittently. These are staggered to optimise internal spaces, and the staggering creates vertical apertures for cross-ventilations, day-lighting, and access to the outdoor and greenery.

The final result brings back fond memories of the couple's childhood days – the days of living in a community where homes were interconnected with social spaces that were simple and adaptive; and with the rituals of everyday life enriched by architecture.

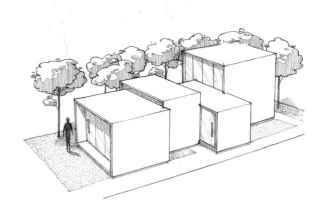

Binh Thanh 住宅
Vo Trong Nghia Architects + Sanuki + Shunri Architects

该住宅位于越南胡志明市中心，是为两个家庭所设计；一对六十多岁的夫妇和其儿子儿媳带着孩子一家。

该住宅场地具备双重特征。它位于城市当中一处典型的发展和城市化区域之中，面向一条喧闹且尘土飞扬的街道。但同时该场地又非常靠近河流，以及拥有大量绿色植物的西贡草禽园。

鉴于其环境的这一双重背景，该住宅的设计理念是构思出热带气候中的两种不同生活方式；其中一种为自然而传统的生活方式，通过水和绿色植物来进行自然采光和通风；而另一种为现代和优质的生活方式，配备有空调这样的机械设备。

该住宅由两个相互错开的不同空间组成。拥有现代生活方式的空间设置在混凝土花纹体块包覆的三个悬浮体量之中。两个体量之间有两个由玻璃门包覆的夹层空间，并且它可以完全打开，面向室外，在那里住户可以享受清风、日光、绿植和水带来的自然生活。

三个体量交错转换，将自然光线带入到夹层空间当中，并在每层都形成一个小型花园。而底层体量便成为夹层空间的天花板。这些表面的曲线设计各不相同，从而使得每一夹层空间的光线效果都各不相同。

卧室和其他小房间都设置在半封闭式的悬浮体量当中，以提高安全性和隐私性。也就是说，开放的夹层空间是用作两个家庭的独立起居空间的。

花纹体块曾是越南十分受欢迎的一种设计，通过使用它可以获取自然通风。这种体块由长60cm、高40cm的预制混凝土制成。它不仅可以阻挡刺目的阳光和暴雨，还能够提高隐私性和安全性。

虽然该住宅与胡志明市的典型联排别墅看起来有所不同，但所有的建筑解决方案都是出自当地的生活方式和智慧。而在该住宅中，和谐共存的现代和自然的生活方式诠释了现代热带城市中的生态生活方式。

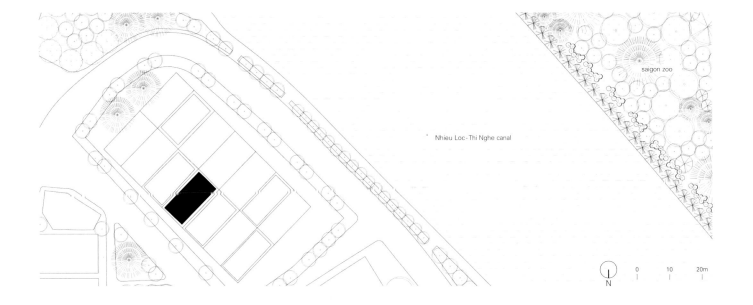

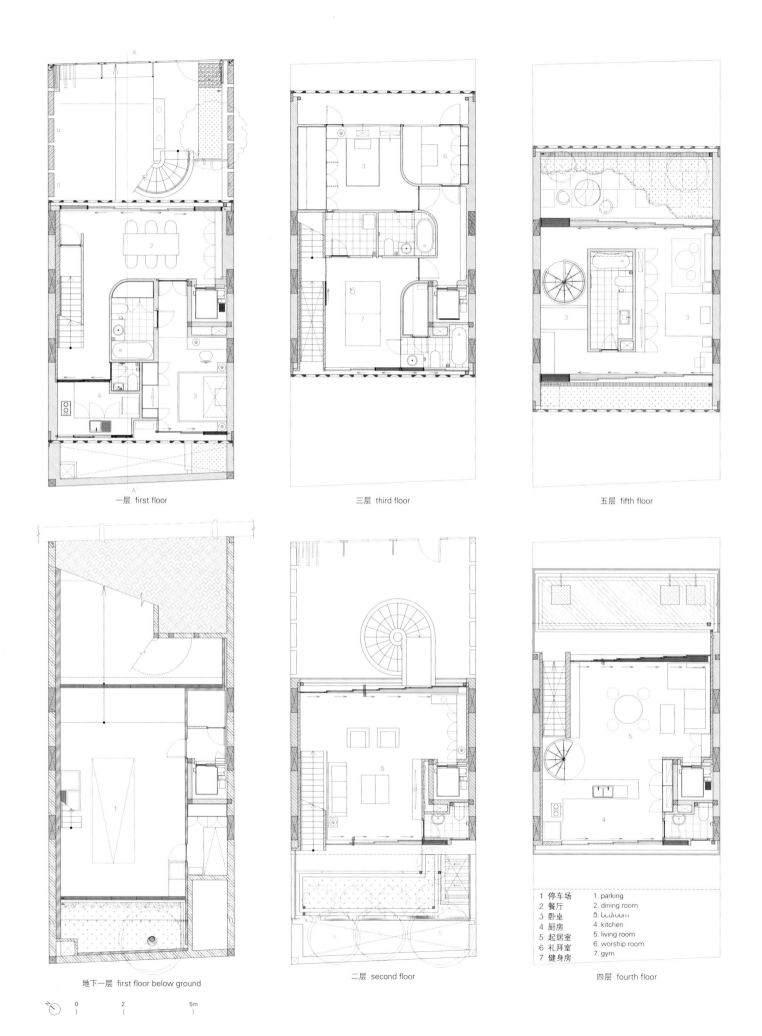

项目名称：Binh Thanh House
地点：Binh Thanh, Ho Chi Minh City, Vietnam
建筑师：Vo Trong Nghia Architects,
Sanuki + Nishizawa Architects
承包商：Wind and Water House JSC, Danang Company
用地面积：140m² 总建筑面积：81m²
有效楼层面积：516m²
竣工时间：2013.6
摄影师：©Hiroyuki Oki(courtesy of the architect)

1 餐厅
2 浴室
3 厨房
4 起居室
5 卧室
6 健身房

1. dining room
2. bathroom
3. kitchen
4. living room
5. bedroom
6. gym

A-A' 剖面图 section A - A'

Binh Thanh House

Located in the center of Ho Chi Minh city in Vietnam, the house was designed for two families; a couple in their sixties and their son's couple with a child.

The plot has a bilateral character. It is in a typical developing and urbanizing area in the city, facing to a noisy and dusty street. But it is also very close to the river and the Saigon Zoo with a plenty of greenery.

Against a backdrop of this duality of its setting, the concept of the house is to accommodate two different lifestyles in a tropical climate; one is a natural and traditional lifestyle, utilizing natural lighting and ventilation with water and greenery, and the other is a modern and well-tempered lifestyle with mechanical equipments such as air-conditioners.

The house is composed of two different spaces positioned alternately. Spaces for modern lifestyle are allocated in three floating volumes wrapped by concrete pattern blocks. Between volumes are two in-between spaces covered by glasses and widely open to the exterior, where the residents enjoy their natural life with wind, sunlight, green and water.

Three volumes are shifted back and forth to bring natural light into the in-between spaces, as well as to create small gardens on each floor. The bottoms of the volumes become the ceilings for the in-between spaces. These surfaces are designed with various curved shapes, providing each in-between space with different light effect.

Bedrooms and other small rooms are contained in the floating semi-closed volumes to enhance security and privacy. On the other hand, the open in-between spaces are designed to be independent living spaces for two families.

Pattern blocks, which used to be a quite popular device in Vietnam to get matural ventilation, are made of pre-cast concrete with 60cm width and 40cm height. It helps a lot not only to prevent the harsh sun-light and heavy rain but also to enhance the privacy and safety.

While this house looks different from the stereotypical townhouses in Ho Chi Minh City, all the architectural solutions are derived from the local lifestyle and wisdom. The house, in which modern life and natural life are compatible with each other, offers an interpretation of the ecological lifestyle in the modern tropical city.

Venturini住宅
Adamo-Faiden

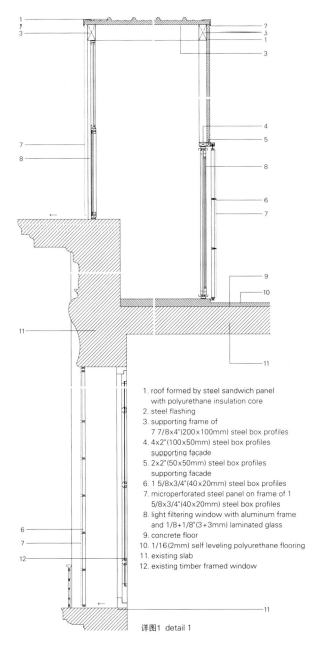

1. roof formed by steel sandwich panel with polyurethane insulation core
2. steel flashing
3. supporting frame of 7 7/8x4"(200x100mm) steel box profiles
4. 4x2"(100x50mm) steel box profiles supporting façade
5. 2x2"(50x50mm) steel box profiles supporting facade
6. 1 5/8x3/4"(40x20mm) steel box profiles
7. microperforated steel panel on frame of 1 5/8x3/4"(40x20mm) steel box profiles
8. light filtering window with aluminum frame and 1/8+1/8"(3+3mm) laminated glass
9. concrete floor
10. 1/16(2mm) self leveling polyurethane flooring
11. existing slab
12. existing timber framed window

详图1 detail 1

该住宅位于Abasto市场的附近，如今已经完全转换为一个商业中心。和市场一样，这处场地是Venturini家族用做不同用途的地方。在施工期间，它成为一间出租房，其布局和布宜诺斯艾利斯市的常规类型相呼应，小房屋面向街区的内部，而业主房子的一侧墙体则充当临街的立面。在20世纪中叶，这一城市部分的贬值使大量占领的主建筑都被改造成出租屋。

建筑师的干预可以归纳为三个步骤：提取、重新加以描述和添加。第一个步骤意味着对原有空间进行修复。而项目的第二阶段以突出每处空间为基础，使现有的结构适应当代的生活方式。最后一个步骤是增建两个精致的结构。其中一个结构为夹层，使地下室能够容纳一间瑜伽室，同时能对起居室的外向扩建提供支撑。最后，一个轻质结构位于屋顶，且正处于施工期间，以用于多功能使用，并且使人们感受到积极向上的氛围，以遵循城市的升值性能，成为一种信息遗产复兴方式。

Venturini House

The house is located close to the Abasto Market, nowadays converted into a commercial center. Like the market, the place where the Venturini family lives homed a variety of different uses. At the time of the construction it functioned as a house of rent. Its organization responded to a very common typology in the city of Buenos Aires. Small houses were located towards the interior of the block, whereas the one belonging to the owner was the facade to the street. The devaluation of this part of the city towards the middle of last century brought about the occupation of the main house, being transformed as a result into a tenement house. The architects' intervention can be summarized into three actions: extraction, re-description and addition. The first of them meant the recovery of the original spatial structure. The second phase of the project was based simply on marking again each of the spaces in order to adapt the existing structure to a contemporary way of life. Finally, the last action consisted in two precise additions. The first of these was the materialization of a mezzanine floor which allowed to simultaneously cover a yoga room in the basement and give support to an exterior expansion for the living room. At last, the construction of a light structure for multiple uses on the roof made visible the optimism that follows the revalue of the city as a way of new crowning for the property. Adamo-Faiden

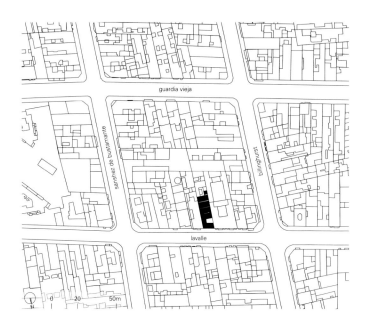

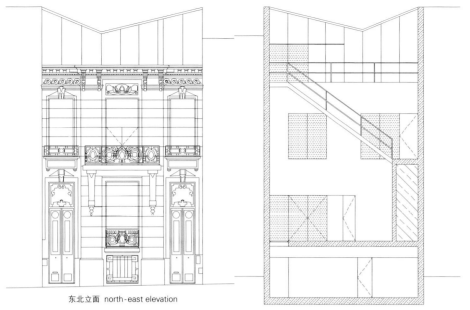

东北立面 north-east elevation

西南立面 south-west elevation

1 卧室	1. bedroom
2 洗衣房	2. laundry
3 衣橱	3. wardrobe
4 体育馆	4. gym
5 入口	5. entrance
6 厨房	6. kitchen
7 大厅	7. hall
8 起居室/餐厅	8. living/dinning room
9 露台	9. patio
10 书房	10. study room
11 平台	11. terrace
12 桑拿室	12. sauna room
13 餐厅	13. dinning room

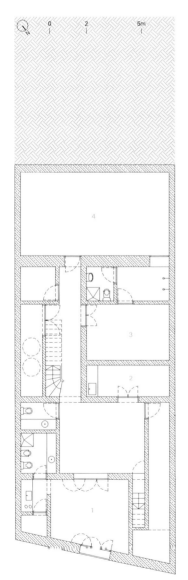

地下一层 first floor below ground

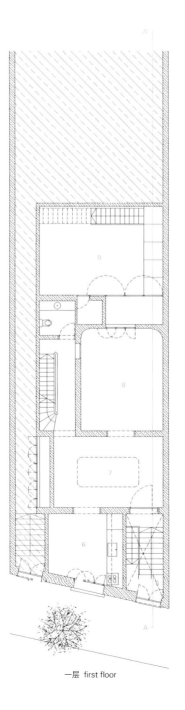

一层 first floor

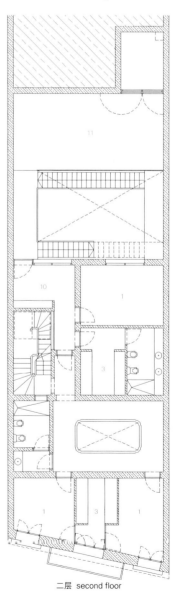

二层 second floor

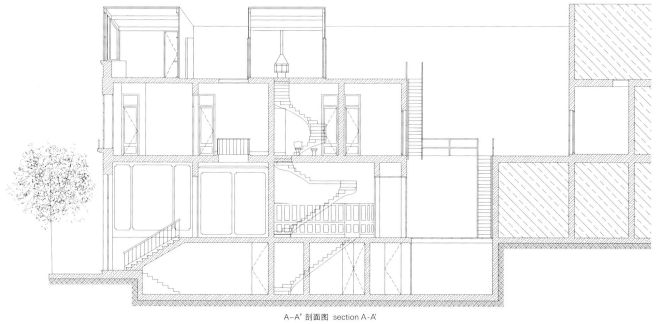

A-A' 剖面图 section A-A'

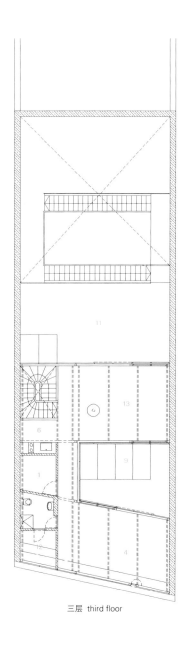

三层 third floor

项目名称：Venturini House
地点：Lavalle 3470, Ciudad Autónoma de Buenos Aires, Argentina
建筑师：Adamo-Faiden
项目团队：José Castro Caldas, Flore Silly, Gabriela Schaer
甲方：Familia Venturini
用地面积：204m² 有效楼层面积：462m²
设计时间：2010 竣工时间：2011
摄影师：©Cristóbal Palma

Alcobaça住宅
Aires Mateus

邻里住宅 Dwell How Neighborhood

在Alcobaça历史中心设计的这座住宅记录着重叠的时间：一座小型建筑进行了重建，以延续当地常规的建筑规模，而一堵墙则完全圈出僻静的扩建区域。在现存的建筑中，上空体量用于调整周边墙体的厚度。建筑内没有设置收集天窗射来的阳光的空间，以形成私密且受保护的氛围。室内设有隔间，作为室内的增建结构，通过立面上的窗户与室外相连。住宅的扩建在两个不同水平高度，即街面和与Baça河共生的花园，为不同的结构。新建墙体围合出庭院，而社交区域延续着空间，在嵌入结构中来回蜿蜒。

N 0 10 30m

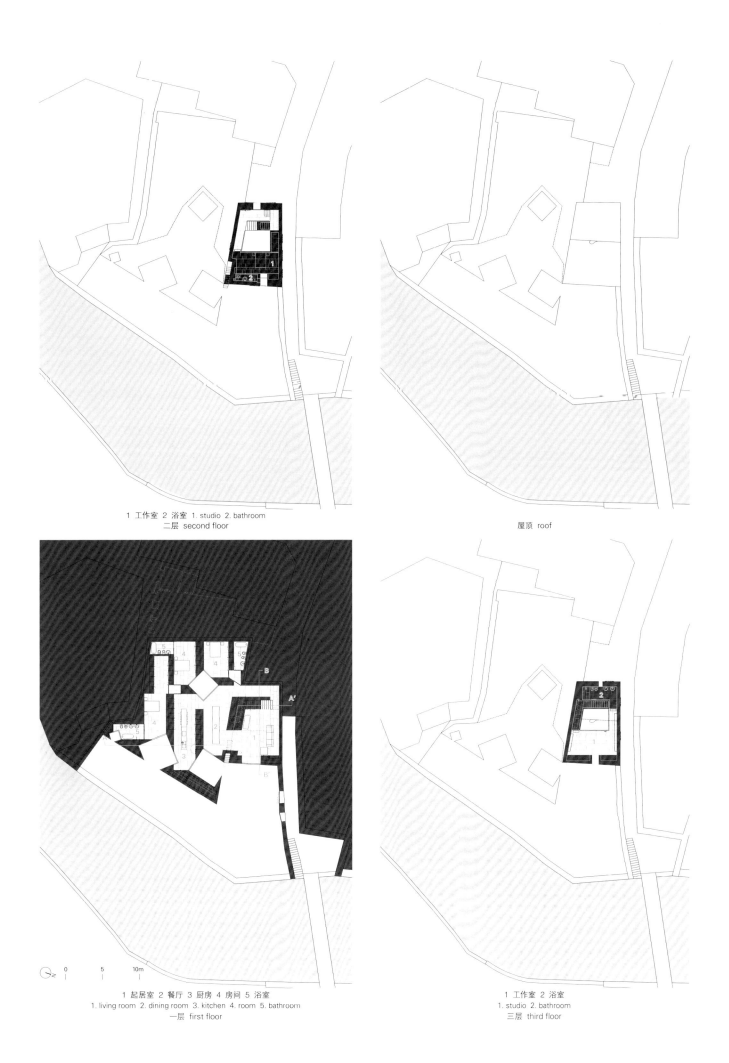

1 工作室 2 浴室 1. studio 2. bathroom
二层 second floor

屋顶 roof

1 起居室 2 餐厅 3 厨房 4 房间 5 浴室
1. living room 2. dining room 3. kitchen 4. room 5. bathroom
一层 first floor

1 工作室 2 浴室
1. studio 2. bathroom
三层 third floor

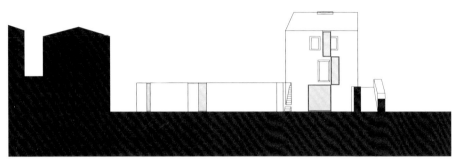

东北立面 north-east elevation

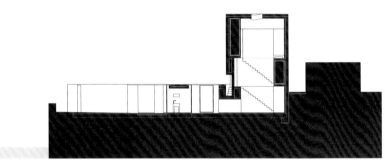

A-A' 剖面图 section A-A'

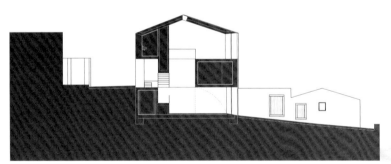

B-B' 剖面图 section B-B'

Alcobaca House

The house designed in the historical center of Alcobaça is a record of overlapping times: A small building reconstructed to perpetuate the vernacular common scale, and a wall thoroughly shaped to house the quiet extension. On the existing building, a void was made to manage the thickness of the peripheral walls. An absence of space is freed for collecting luminosity from a skylight that grants a private and protected atmosphere. The compartments appear as internal additions, connected to the exterior through windows on the facades. The extension of the house takes the difference between two levels: The street level and the garden that is generated with the River Baça. The form of the new wall defines courtyards that mediate the contemplation to the exterior. The social areas, work as a spatial continuum that spread through the two times of the intervention.

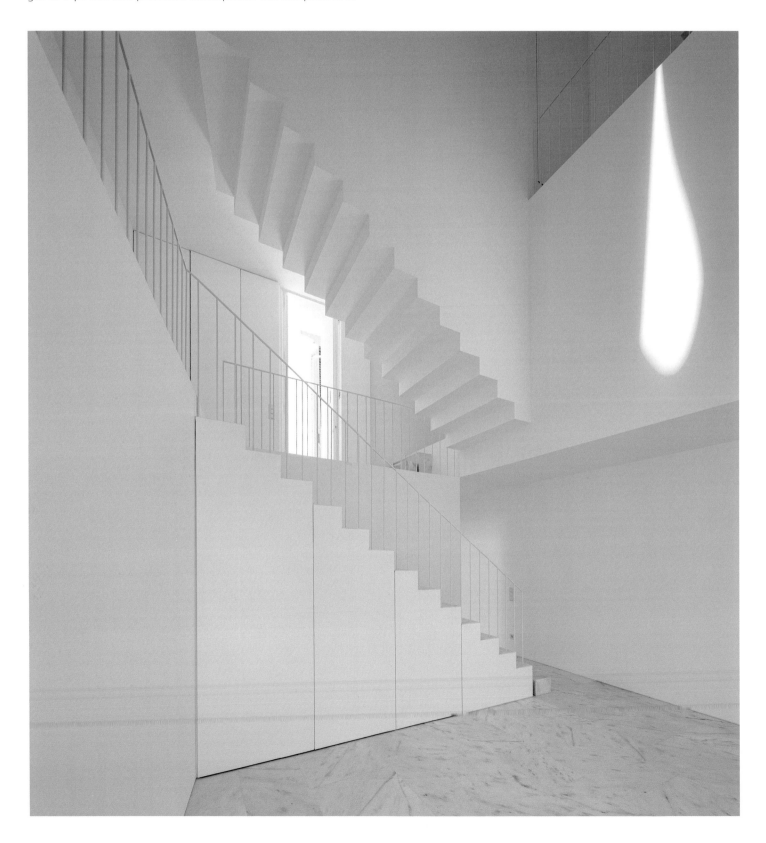

项目名称：House in Alcobaça
地点：Alcobaça, Portugal
建筑师：Aires Mateus
合作商：Catarina Bello
工程师：Betar, Ecoserviços
建造商：Manuel Mateus Frazão
施工协调：aime Coelho
面积：475m²
施工时间：2010—2011
摄影师：©FG+SG Architectural Photography

103住宅
Marlene Uldschmidt Architects

居住在墙体后面

对于接到葡萄牙南部Arade河宽阔的河口处的Ferragudo区渔村内的一个如此有趣的项目，Marlene Uldschmidt建筑师事务所表现得十分的兴奋。Ferragudo的历史中心是一处极其敏感的工作区域，但是建筑师相信他们的增建建筑能够在此地和谐共处，且与周围的建筑和历史融为一体。该工作室决定深度开发"居住在墙体后面"这个理念。

建筑师所面临的挑战是建造一处立面，使其成为公共区域和私人区域之间的一道实体屏障，同时能够强化与村庄和河流的视觉连接。

场地较难开发的地形意味着这个理念还要允许对室内进行设计，以突出与村庄其他区域甚至以外的区域的视觉连接。因此，建筑师选择的理念主要用来在场地内改变高度，以达成这一目标。

该理念所提出的另一个挑战是在室内形成一种明亮且通风的感觉。建筑师建造了一个垂直采光井，将所有楼层连接起来，以完成这一目标。

为了平衡白色墙体的朴素感，木质自然材料和土色石材也应用到建筑内。

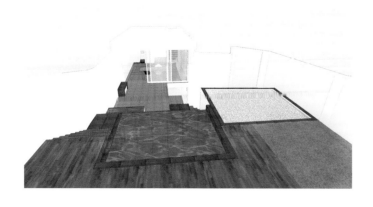

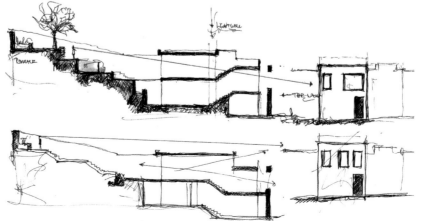

103 House
Living behind the Wall

Marlene Uldschmidt Architects was excited to take on such an interesting project in the Fishing Village of Ferragudo, which is situated on the wide estuary of the River Arade in the south of Portugal. The historic center of Ferragudo is an extremely sensitive area to work in and they believed that their intervention should be balanced harmoniously and above all integrated with the surrounding architecture and history. The studio decided to explore the concept of "Living behind the wall".

The challenge was to create a facade which would be a physical barrier between the public and private areas whilst enhancing the visual connection with the village and the river.

The difficult topography of the site meant that the concept would need to allow for the design of the internal space to strengthen their visual connection with the rest of the village and beyond. The concept the architects chose was to use the changes of level within the site in order to achieve this goal.

Another challenge of this concept was to create a light and airy feeling within the building. The architects created a vertical well of light that links all levels to achieve this.

In order to balance the simple white walls, natural materials of wood and stone in earthy tones were chosen.

Marlene Uldschmidt Architects

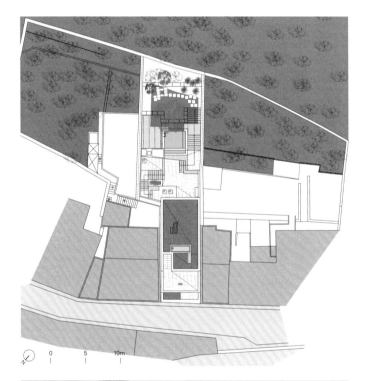

一层 first floor　　　　二层 second floor　　　　三层 third floor

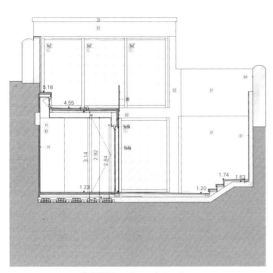

东南立面 south-east elevation

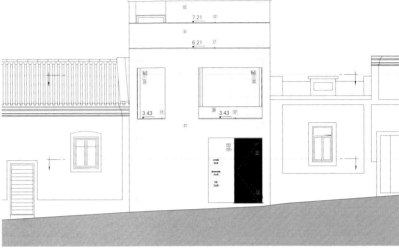

西北立面 north-west elevation

170

项目名称：Casa 103
地点：Ferragudo, Algarve, Portugal
建筑师：Marlene Uldschmidt Architects
合作商：Maurícia Bento
结构工程师：Protecna
照明工程师：LuzArq
木匠：Equipa Quatro
卫生设备设计：JRB
百叶系统：T.S.Pinto, Lda
用地面积：230m²
总建筑面积：90m²
有效楼层面积：180m²
设计时间：2010
竣工时间：2013.5
摄影师：©FG+SG Architectural Photography

172

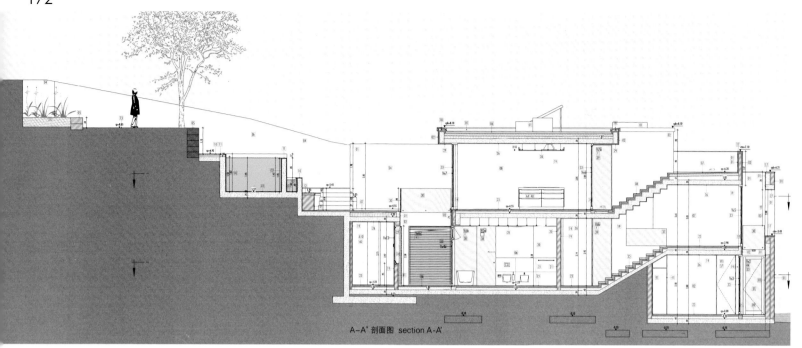

A-A'剖面图 section A-A'

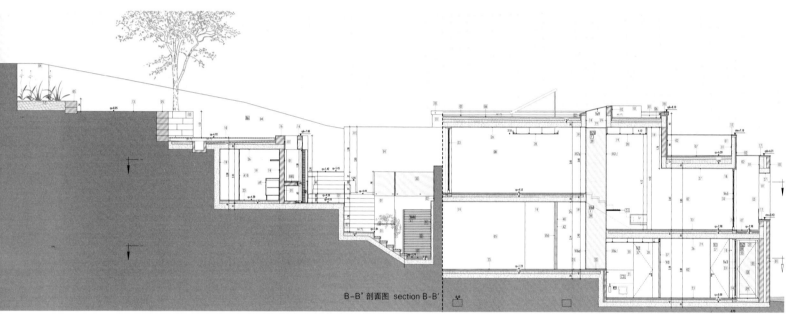

B-B'剖面图 section B-B'

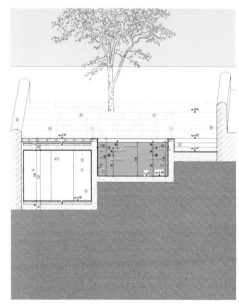

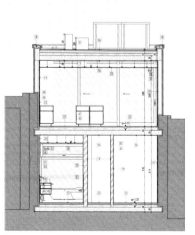

C-C'剖面图 section C-C'   D-D'剖面图 section D-D'   E-E'剖面图 section E-E'

	Paredes exteriores	12	Degraus do jardim em deck de madeira		Pavimentos interiores		Vãos exteriores
01	Revestimento em reboco hidrófugo areado fino pintado com branco cor de cal	13	Terra vegetal com espécies fornecidas e colocadas pela empresa da especialidade	23	Revestimento em pedra natural azul valverde com acabamento amaciado mate conforme esquema de colocação, com espessura de 2 cm e dimensões: 8x49cm; 32x49cm e 49x49cm	32	Porta de entrada com uma folha de abrir para o interior em madeira escurecida conforme cor dos degraus interiores (ver mapa de vãos)
02	Revestimento em Capoto pintado com branco cor de cal		Outros materiais exteriores			33	Janelas com caixilharia em madeira pintada de branco com vidro duplo de 6+6mm com lâmina de ar de 12mm (ver mapa de vãos)
03	Revestimento da piscina e tanque de compensação em pedra natural a definir	14	Peitoris/soleiras em pedra natural Azul valverde com 3 cm de espessura	24	Revestimento em seixo rolado no poço de luz com tons brancos e tamanho a definir	34	Clarabóia com vidro duplo de 6+6mm com lâmina de ar de 12mm (ver mapa de vãos)
04	Muro existente a recuperar, com acabamento rústico, pintado com branco cor de cal	15	Soleira de entrada em pedra natural Azul valverde com 3 cm de espessura	25	Degraus forrados com peças em madeira maciça escurecida, cobertor com 121x30x7cm e espelho com 121x17,5x5cm	35	Portadas exteriores de ensombramento em ripado de madeira no quarto principal (ver mapa de vãos)
05	Muro em blocos de pedra	16	Capeamento em pedra natural azul valverde com 2 cm de espessura		Tectos	36	Portadas de áreas técnicas em ripado de madeira (ver mapa de vãos)
	Coberturas não acessíveis	17	Capeamento dos vãos da fachada e platibanda em mármore branco	26	Revestimento interior em estuque pintado na cor branca		Vãos interiores
06	Revestimento de cobertura plana não acessível em seixo rolado com tamanho e cor a definir	18	Rufo em zinco	27	Revestimento exterior em reboco hidrófugo areado fino pintado na cor branco	37	Portas de madeira pintadas de branco (ver mapa de vãos)
	Pavimentos exteriores		Paredes interiores	28	Tecto falso em placas de gesso cartonado pintado na cor branco	38	Porta de vidro de correr em conjunto com vidro fixo (ver mapa de vãos)
07	Revestimento em pedra natural azul valverde com acabamento amaciado mate, estereotomia conforme esquema de colocação, com espessura de 2 cm e dimensões: 8x49cm; 32x49cm e 49x49cm	19	Revestimento em estuque pintado na cor branco	29	Revestimento exterior em Capoto pintado com branco cor de cal	39	Coluna de vidro duplo de 6+6mm com lâmina de ar de 12mm (ver mapa de vãos)
08	Degraus de acesso ao terraço em pedra natural azul valverde com acabamento bojardado de 30x1,20x2cm	20	Revestimento com reboco hidrófugo areado fino nas instalações sanitárias		Guardas exteriores		Armários
09	Revestimento do pátio da entrada em seixo rolado, com tamanho e cor a definir	21	Revestimento com pedra azulvalverde amaciada mate com 2 cm de espessura, 60 cm de altura, e larguras variáveis conforme desenhos parciais das instalações sanitárias	30	Guarda em vidro temperado incolor com prolongadores em aço	40	A _ Armário em madeira (ver mapa de armários)
10	Deck de madeira na área da piscina				Guardas interiores	41	R _ Roupeiros em madeira (ver mapa de roupeiros)
11	Bordadura da piscina e tanque de compensação em pedra natural a definir	22	Revestimento em vidro leitoso com a superfície interior pintada na cor branco por cima do balcão da lava loiça na cozinha, com 30 cm de altura e 2 m de largura	31	Guarda em vidro temperado incolor com prolongadores em aço		

Flynn马厩改建住宅
Lorcan O'Herlihy Architects

邻里住宅

在都柏林的中心地区，这座独栋马厩改建的住宅利用格鲁吉亚古老的场地，与现代美学完美地融合起来。这座住宅采用了1847年建造的马厩房的立面，立面经过整修以及最小幅度的调整。马厩房和其主要场地之间的视觉联系仍然还在，而这些参数形成一个透明且有诚意的设计方法，来为Flynn马厩改建住宅的设计根基提供帮助。

人们从巷子里进入到前院，便会看见住宅的前面由板状的彩色混凝土和玻璃构成，设有一条白石膏点缀的入口通道。逐渐向下走，通道变换成漏斗形，引导客人穿过最初的体量，进入一座封闭的错层式花园。在庭院里，马厩房的立面反射在入口的玻璃幕墙，以及将入口与场地古旧区域连接起来的现代化桥体上面。

作为都柏林绿色建筑试点项目中的一部分，这个项目将一系列重要的、通过整体设计方法来实现的可持续性措施结合在一起。太阳能板用于室内水加热，而辐射采暖地面则利用结合灰水设备的地下热泵系统来工作。住宅所采用的材料包括彩色混凝土、可回收玻璃、高性能保温玻璃以及高光泽的石膏。

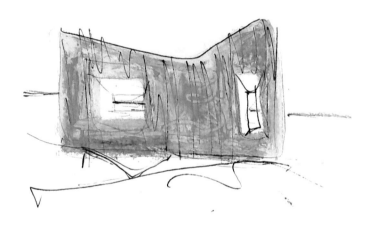

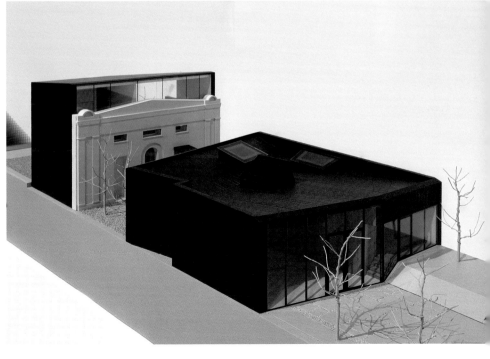

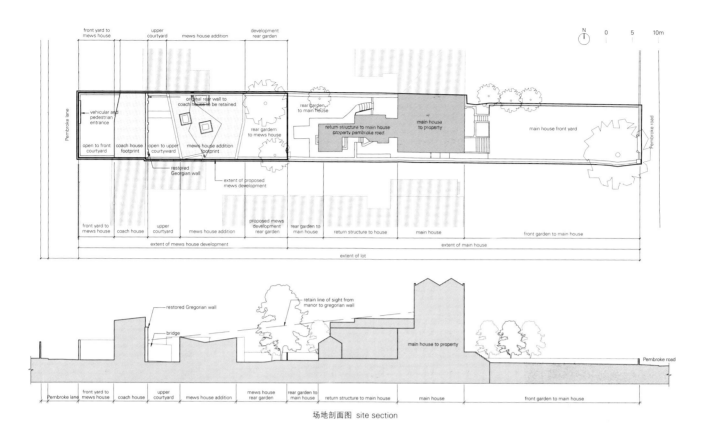

场地剖面图 site section

Flynn Mews House

In the heart of Dublin, this single-family mews marries modern aesthetics with its historic Georgian site. The home incorporates an 1847 coach house facade, which was restored and minimally altered. The visual link between the coach house and its primary manor has, too, been maintained; these parameters drove a transparent and honest design approach that pays homage to the Flynn Mews House's origins.

Entering from the alley into the forecourt, the home's front face is a composition of board-formed stained concrete and glass, with an entry passage highlighted by white plaster. Gradually sloping downward, the passageway funnels the guest through this initial volume and into an enclosed split-level garden. Here in the courtyard, the coach house facade is reflected upon the curtain-wall glazing of the entrance form and the contemporary bridge that joins it with the site's older half.

As part of the Dublin Green Building Pilot Program, the project incorporates a significant amount of sustainable measures achieved through a holistic design approach. Solar panels are used for domestic water heating and radiant floors utilize an underground heat pump system that incorporates gray water. Materials include stained concrete with recycled glass content, high performance insulated glass, and high gloss plaster.

项目名称：Flynn Mews House　地点：Dublin, Ireland
建筑师：Lorcan O'Herlihy Architects
项目总监：Donnie Schmidt　项目经理：Alex Morassut
执行建筑师：ODOS Architects
结构工程师：Casey O'Rourke & Associates
景观建筑师：James Doyle & Associates
甲方：Ella Flynn　承包商：Oikos Builders
功能：four bedrooms for single family
用地面积：415m²　总建筑面积：200m²　有效楼层面积：290m²
设计时间：2009　竣工时间：2012
摄影师：
©Enda Cavanagh (courtesy of the architect) - p.174, p.177, p.178
©Alice Clancy (courtesy of the architect) - p.176, p.179

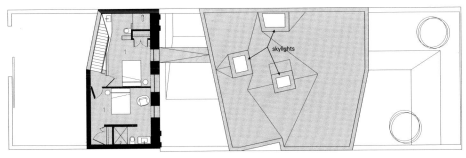

1 卧室　1. bedroom
二层　second floor

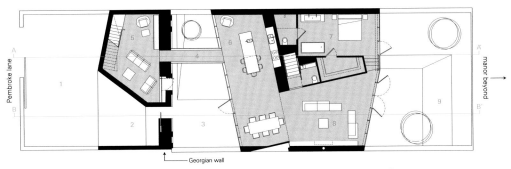

1 入口庭院　2 通道　3 庭院　4 桥　5 媒体间　6 厨房　7 主卧室　8 起居室　9 后花园
1. entry court 2. passageway 3. courtyard 4. bridge 5. media room 6. kitchen 7. master bedroom 8. living room 9. rear garden
一层　first floor

1 卧室　2 洗衣房　3 低层庭院
1. bedroom 2. laundry 3. lower courtyard
地下一层　first floor below ground

1 入口庭院　2 卧室　3 媒体间　4 桥　5 厨房　6 低层庭院　7 楼梯　8 主卧室　9 后花园
1. entry court 2. bedroom 3. media room 4. bridge 5. kitchen 6. lower courtyard 7. stair 8. master bedroom 9. rear garden
A-A' 剖面图　section A-A'

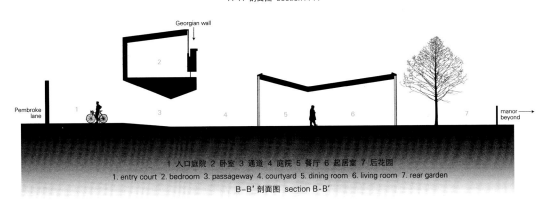

1 入口庭院　2 卧室　3 通道　4 庭院　5 餐厅　6 起居室　7 后花园
1. entry court 2. bedroom 3. passageway 4. courtyard 5. dining room 6. living room 7. rear garden
B-B' 剖面图　section B-B'

>>88
Future Architecture Thinking
Is an internationalized company developing projects from Lisbon and over 25 countries and to the entire world. In addition to Architecture and Urban Planning, they perform Landscape Architecture and Design projects.
Miguel Correia was born in Lisbon in 1961. In 1987, he earned a degree in Architecture from the School of Fine Arts at Lisbon University. He has worked as CEO at IDF from 1987 to 2011. He is also a member of Portuguese Professional Architecture Society.

>>152
Aires Mateus
Manuel Aires Mateus was born in 1963 and Francisco Aires Mateus was born in 1964. They both graduated from Faculty of Architecture at Technical University of Lisbon. Started working together as Aires Mateus Arquitectos from 1988. Manuel has taught at several universities and currently is a professor of Autonomous University and Lusíada University. Francisco was a visiting professor at Oslo School of Architecture and a teacher in Graduate School of Design, Harvard University in 2005. Now he teaches at Autonomous University in Lisbon.

>>146
Adamo-Faiden
Sebastián Adamo and Marcelo Faiden are architects graduated from the University of Buenos Aires and are Ph.D. candidates at the Barcelona School of Architecture. They work as associates, and teach project design at the University of Buenos Aires, the University of Palermo and the University Torcuato Di Tella. They have exhibited their work in the Guggenheim Museum in New York, in the Argentine and British Pavilions of the Venice Architecture Biennale and in Mexico City at LIGA Space for Architecture. In the XII International Architecture Biennale in Buenos Aires, they were awarded with the gold medal.

>>6
Steven Holl Architects
Was founded in New York in 1976 and has offices in New York and Beijing. Steven Holl leads the office with partners Chris Mcvoy and Li Hu. Graduated from the university of Washington and pursued architecture studies in Rome in 1970. Joined the Architectural Association in London in 1976. Is recognized for his ability to blend space and light with great contextual sensitivity and to utilize the unique qualities of each project to create a concept-driven design. Specializes in seamlessly integrating new projects into contexts with particular cultural and historic importance. Is a tenured faculty member at Columbia University.

>>66
A+ Architecture
Is a professional design practice consisting primarily of architects, and designers, graphic designers and urban planners. It has its references in various public facilities, office buildings and offices, private and public housing. Philippe Bonon[left] graduated from National and Superior School of Beaux-Arts, Paris in 1983. In 1995, he created the BBA Ltd. with partner, Denis Bedeau. Philippe Cervantes[middle] graduated with a major in building at the ESTP Paris in 1986. In 2001, he became partner and co-director of BBA Ltd. Gilles Gal[right] is project manager and construction economist. He is in charge of the management of construction sites and manages the design studies and prescription.

>>128
Chang Architects
Born and raised in Singapore, Yong Ter's passion for architecture was discovered during his university years at the School of Architecture, National University of Singapore. Upon graduation, he sought apprenticeship with Mr. Tang Guan Bee for several years, before starting his practice, Chang Architects, at the turn of this millennium. In his early years of practice, he was one of the 20 architects to be selected by the Urban Redevelopment Authority of Singapore to showcase the works of young and emerging architects in Singapore. In recent years, the practice has received several architectural awards.

>>76
Artech Architects
Kris Yao obtained a B.Arch from Tunghai University in 1975, and a M.Arch from University of California, Berkeley in 1978. He established Artech Architects in Taipei, Taiwan, China in 1985, and Artech Architects & Designers Limited in Shanghai, China in 2001. In 1997, he was awarded the 3rd Annual "Chinese Outstanding Architect Award". In 2002, he represented Taiwan in the 8th International Architecture Exhibition at La Biennale di Venezia, Italy. In 2007, he became the first practicing architect to receive the "National Award for Arts" in Architecture, the highest honor in cultural and art disciplines in Taiwan.

Andrew Tang
Received his architecture degree at Institute of Design(IIT), Chicago and won the Jerrold Loebl Prize in 1996. Has worked around the world with many prominent figures and reputable offices such as West 8, MVRDV, Maxwan, and Architecten Cie before starting his own design practice, Tanglobe. Has also written many articles and lectured in many institutions and conferences with a strong passion for architecture.

Heidi Saarinen
Is a senior lecturer in Interior Architecture and Design at the University of Hertfordshire, and also external advisor at University of the Arts London and external examiner at Coventry University, UK. Was born in Finland, raised in Sweden, and is now based in London. Has extensive teaching experience and project management, with specialism in experiential methodologies and spatial analysis in teaching and learning, encouraging students to interact with space through movement; questioning behavior and use of space. Is currently working on collaborative projects linking film, digital media, architecture and design, community and narrative space.

Marta González Antón
Is an architect, works and lives in London. Studied in Spain and Italy. Before moving to London she collaborated with several local offices in Spain and worked on publication activities. Is currently carrying out a project research about the contemporary design process between the Netherlands and Spain.

>>16
MTM Arquitectos
Javier Sanjuan Calle and Javier Fresneda Puerto both graduated from the Polytechnic University of Madrid in 1992. They have been working as associates of MTM Arquitectos since 1997. Among the 17 built works, various congratulations through prizes and selections were given and they have obtained more than 37 prizes in competitions.
Javier Sanjuan Calle is a specialist in urban planning. He is teaching Architectural Design at the European University of Madrid. Javier Fresneda Puerto is a specialist in structure. He is also teaching Architectural Design at the Architecture School Alcala de Henares.

>>138
Vo Trong Nghia Architects + Sanuki + Nishizawa Architects
Vo Trong Nghia graduated from Nagoya Institute of Technology with a B.Arch in 2002 and received Master of Civil Engineering from Tokyo University in 2004. In 2006, he established Vo Trong Nghia Co.Ltd. Daisuke Sanuki received a B.Arch in 1998 and M.Arch in 2000 from Tokyo University of Science. Was a partner of Vo Trong Nghia Co.Ltd. from 2009 to 2011 and established S+Na Co.Ltd. In 2011. Shunri Nishizawa received a B.Arch in 2003 and M.Arch in 2005 from Tokyo University. Has worked at Tadao Ando Architects and Associates and was a partner of Vo Trong Nghia Co.Ltd. between 2009 and 2011. Currently runs S+Na Co.Ltd. with Daisuke Sanuki.

>>106
Yuko Nagayama & Associates
Yuko Nagayama was born in Tokyo in 1975. Graduated from Showa Women's University in 1998 and established Yuko Nagayama & Associates in 2002 after working at Jun Aoki & Associates. Received numerous awards including Architectural Record Award, Design Vanguard 2012. She is teaching at Kyoto Seika University, Showa Women's University, Ochanomizu University, and Nagoya Institute of Technology. Representative work includes Louis Vuitton Kyoto Daimaru, Anteprima Singapore, Kiya Ryokan etc. Continues to participate on various exhibitions like Kenchiku×Architecture and solo exhibition "Exhibition of Yuko Nagayama".

>>164
Marlene Uldschmidt Architects
Marlene Uldschmidt studied at Academy of Fine Arts and Deutscher Werkbund in Nuernberg, Germany from 1985 to 1987. From 1987 to 1992, she studied architecture and design at the University of Hildesheim in Hannover, Germany. Has created and operated sTUDIO – aTELIER Marlene Uldschmidt since 2004 developing architectural concepts and design solutions. In 2011, she launched Ultramarino offering interior design and project management services.

>>116
Enota
Was founded in 1998 with the ambition to create contemporary and critical architectural practice of an open type based on collective approach to development of architectural and urban solutions. The main focus is research driven design of the environment where contemporary social organizations, new technologies and arts are interwoven. Milan Tomac and Dean Lah have been partner architects of Enota since 2002.

>>174
Lorcan O'Herlihy Architects
Is committed to engage the complexities of contemporary society through architecture. Approach their work with ruthless optimism, dedicated to a conviction that architecture can awaken people and enrich communities. Their process is collaborative and iterative. Understand that their architecture is accomplished by and for people. Since 1990, LOHA has built over 75 projects across three continents. Their work ranges in typology from institutional buildings to bus shelters, and from large-scale developments to single-family homes. Has been published in over 20 countries and recognized with over 100 awards, including the 2010 AIA Los Angeles Firm of the Year.

>>94
Daipu Architects
Dai pu was born in 1983 and studied 5 years in Huazhong University of Science & Technology and received a Bachelor of Architecture. Worked as an assistant architect in NODE(Nansha Original DEsign) at Guangzhou from 2006 to 2007. From 2007 to 2009, he worked as project architect in Limited Design and MAD in Beijing. In 2010, he founded his own practice Daipu Architects. In 2013, he participated in the exhibition "Palace Of China, Architecture of China" at Segovia, Spain and "West Bund, Biennial of Architecture and Contemporary Art" at Shanghai, China.

>>56
Studio Sumo
Yolande Daniels[left] and Sunil Bald[right] have been leading Studio Sumo since 1998. Has received Young Architects awards from the Architectural League of New York and Design Awards from the AIA. Yolande Daniels received a B.Arch from City College and a M.Arch from Columbia University. She has taught over ten years in prestigious academic institutions and working as professor of practice at MIT since 2012. Sunil Bald is a registered architect in the state of New York. He received a M.Arch from Columbia University where he was awarded the AIA Medal. He has taught at Cornell, Columbia, University of Michigan, and Parsons.

>>38
Francis-Jones Morehen Thorp
Intentionally work at various scales, and regularly undertake major consultancies in master planning, exhibition design and community consultation. Design Director, Richard Francis-Jones graduated from the University of Sydney with University Medal for Architecture. Received Master of architectural design and theory at Columbia University in New York. Was President of the AIA (NSW Chapter) in 2001 and 2002.
Managing Director, Jeff Morehen graduated with Honours from Deakin University, and has over twenty years professional experience gained throughout Australia and abroad. Is fully involved with projects as a Project Director for many of the firm's award-winning commissions.

Archimedia
Is an award-winning, New Zealand architecture practice, offering professional design services in the disciplines of architecture, interiors and ecology. They understand that delivering architecture involves intervention in both natural eco-systems and the built environment – the context within which human beings live their lives.
Lindsay Mackie graduated from the University of Auckland with bachelor's degree of Architecture in 1976. She worked as principal of Archimedia in 1998. Hamish Cameron is a practice director who also graduated from the University of Auckland with bachelor's degree of Architecture in 1981. He joined Archimedia in 1988 and has been working as director since 1996.

>>48
Speech Tchoban & Kuznetsov
The bureau Speech was founded in 2006 as a result of long term collaboration of Berlin office "NPS Tchoban Voss" and Moscow office "Tchoban and Partners" directed by Sergei Tchoban[right] with the bureau "S.P. Project", headed by Sergey Kuznetsov[left]. Sergey Kuznetsov was born in 1977 in Moscow and graduated from the Moscow Architectural Institute in 2001. In 2012, he was assigned to the post of the chief architect of Moscow. Sergey Tchoban has been working in Germany since the beginning of 1990s after architecture education in Saint Petersburg. He has been run-

C3, Issue 2013.12
All Rights Reserved. Authorized translation from the Korean-English language edition published by C3 Publishing Co., Seoul.

© 2014大连理工大学出版社
著作权合同登记06-2014年第25号

版权所有·侵权必究

图书在版编目(CIP)数据

体验与感受：艺术画廊与剧院：汉英对照 / 韩国C3出版公社编；王凤霞等译. — 大连：大连理工大学出版社，2014.3
（C3建筑立场系列丛书；36）
ISBN 978-7-5611-8914-6

Ⅰ. ①体… Ⅱ. ①韩… ②王… Ⅲ. ①画廊－建筑设计－汉、英②剧院－建筑设计－汉、英 Ⅳ. ①TU242

中国版本图书馆CIP数据核字(2014)第031600号

出版发行：大连理工大学出版社
　　　　　（地址：大连市软件园路80号　邮编：116023）
印　　　刷：上海锦良印刷厂
幅面尺寸：225mm×300mm
印　　张：11.5
出版时间：2014年3月第1版
印刷时间：2014年3月第1次印刷
出 版 人：金英伟
统　　筹：房　磊
责任编辑：张昕焱
封面设计：王志峰
责任校对：赵姗姗

书　　号：ISBN 978-7-5611-8914-6
定　　价：228.00元

发　行：0411-84708842
传　真：0411-84701466
E-mail：12282980@qq.com
URL：http://www.dutp.cn